Electronics
Build and Learn

Electronics
Build and Learn
Second Edition

R A Penfold

PC Publishing

PC Publishing
22 Clifton Road
London N3 2AR

Second edition 1988

© PC Publishing

First published by Newnes Technical Books 1980

ISBN 1 870775 15 5

All rights reserved. No part of this publication may be reproduced or transmitted in any form, including photocopying and recording, without the written permission of the copyright holder, application for which should be addressed to the Publishers. Such written permission must also be obtained before any part of this publication is stored in a retrieval system of any nature.

The book is sold subject to the Standard Conditions of Sale of Net Books and may not be resold in the UK below the net price given by the Publishers in their current price list.

British Library Cataloguing in Publication Data

Penfold, R.A.
 Electronics: Build and Learn—2nd ed.
 1. Electronics equipment
 I. Title
 621.381

ISBN 1—870775—15—5

Cover design by Tony Walling
Printed and bound by LR Printing Services, Manor Royal, Crawley, West Sussex

Preface

Although it is possible to manage with little or no theoretical knowledge, most electronics enthusiasts want to master the basic theory of the subject at an early stage. This makes it easier to try more complex projects, to be better at fault finding, and to design and adapt circuits to suit one's own needs.

The purpose of this book is to help the complete beginner to understand what the main electronic components do, and how they are used in practical circuits. The first chapter describes a 'circuit demonstrator unit', which can be used to set up quickly and easily many of the experiments and circuits featured in the rest of the book. Thus the practical aspects of electronics are not ignored, and subjects such as soldering and resistor colour codes are covered. It is not essential to build the circuit demonstrator unit and try out the various circuits in order to follow this electronics course. However, most readers will find it beneficial to do so, with the subject matter being more easily understood and remembered as a result.

Although in a book of this length it is only possible to cover some of the simpler theory, it will enable you to understand the operation of all but the more complex designs featured in electronics books and magazines. It should also make it easier for you to extend your knowledge of electronics, and keep up with modern developments.

Most of the components used in the experiments and circuits are popular types that should be of use once the experiments have been completed. Similarly, the circuit demonstrator unit will be useful for testing and experimenting with further simple circuits and projects.

In this revised version of the book some changes have been made to the circuit demonstrator unit, which now uses more modern and more easily obtained components. Also, with integrated circuits (and digital devices in particular) playing an increasingly dominant role in electronics, the new edition has been enlarged to give better coverage of this aspect of the subject. Although modern circuits are dominated by integrated circuits, it is worth stressing the importance of knowing the fundamentals of the basic electronic components. None of the original material covering these aspects has been removed from the current edition.

RAP

Contents

1. **Circuit demonstrator unit**   1
   How to build the unit used for the experiments in Chapters 2 to 6

2. **Passive components**   11
   Conductors and insulators — resistors and potentiometers — capacitors — inductors — transformers

3. **Semiconductor devices**   38
   Diodes, zener diodes and LEDs — bipolar transistors — thyristors and triacs — photodiodes and phototransistors — field-effect transistors

4. **Operational amplifiers**   68
   Voltage comparator — inverting amplifier — non-inverting amplifier

5. **Oscillator and radio circuits**   77
   Relaxation oscillator — phase-shift oscillator — astable multivibrator — tuned-circuit oscillator — two simple radios

6. **Pulse and logic circuits**   91
   Bistable multivibrator — monostable multivibrator — Schmitt trigger — gates — binary — BCD — hex — counters — supply decoupling

Index   109

1

Circuit demonstrator unit

The circuit demonstrator unit shown in Figures 1.1 and 1.2 is based on a piece of good-quality hardboard or thin plywood measuring 200 X 216 mm. The various components (see the list on page 10) are mounted on this using the layout shown in Figure 1.3. The pattern of holes in the top left-hand corner forms a speaker grille; the diameter of

Figure 1.1 The completed unit

1

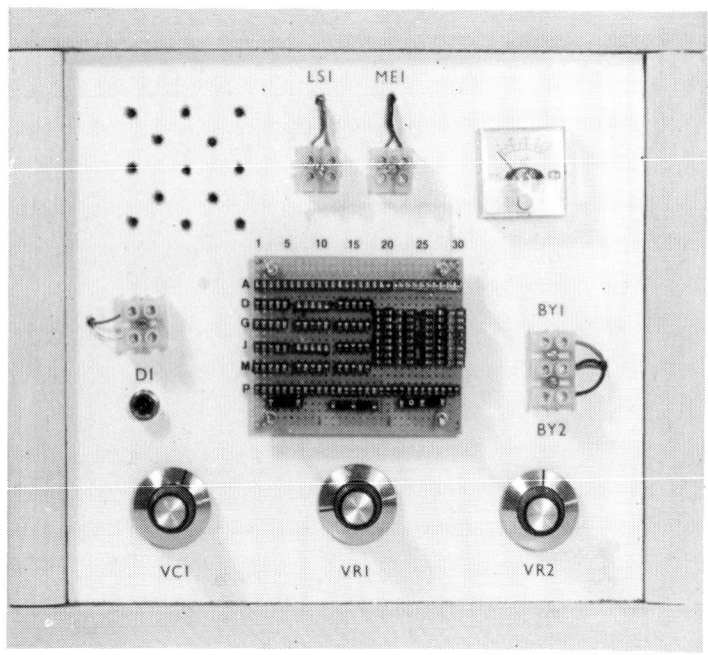

Figure 1.2 Top view of circuit demonstrator unit

these holes can be from 3 to 5 mm. It is unusual for miniature loudspeakers to have provision for screw fixing, so it will almost certainly be necessary to glue this component in position on the underside of the board beneath the grille. It is probably best to leave this until all the drilling of the top panel has been completed. Use only a small amount of a good-quality adhesive, applying it carefully to the outer front rim of the speaker. Avoid smearing adhesive on to the diaphragm, as this could impede its movement and impair the speaker's performance.

Meter mounting requirements vary, and those shown in Figure 1.3 for ME1 are suitable for the standard type sold by most small retailers. However, check this point before making the mounting holes. A small and inexpensive meter is perfectly suitable for the present application. All meters seem to require a large main mounting hole, and this can be made using a fretsaw, coping saw, or a miniature round file. The positions of the smaller holes, where they are needed, can be located using the meter as a sort of template.

A couple of two-way connector blocks are mounted between the meter and the loudspeaker. Each block requires a 6BA or M3 mounting bolt about 25 mm long, plus a fixing nut. A connector block is simply two interconnected screw terminals that can be used to join two or more wires. Two-way blocks have two such sets of terminals. In this

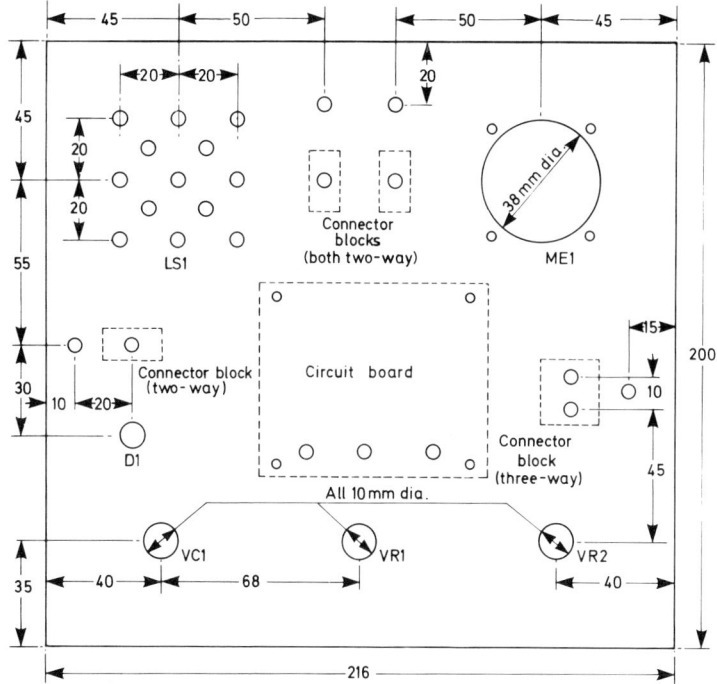

Figure 1.3 Drilling and cutting details for top panel

unit the meter and the loudspeaker each connect to one side of their respective blocks, and leads are connected to the other sides of the blocks when these components are to be connected into a circuit. This avoids having unconnected leads getting in the way when some of the components on the demonstrator are unused. The same method is used to connect the indicator lamp (D1) and the two batteries into circuit when they are required. Note that the batteries share a three-way terminal block, whereas the other three components have individual two-way blocks.

Terminal blocks are only readily available in twelve-way strips, and blocks of the required lengths must be cut down from these using a sharp modelling knife to sever the plastic insulating material that joins the blocks. A hole of about 3–5 mm diameter is drilled beside each terminal block through the hardboard or plywood panel; these holes will eventually accommodate connecting wires.

D1 should be obtained with a panel mounting clip, and the diameter of the mounting hole for this component must be chosen to suit the particular type used.

VC1, VR1, and VR2 are mounted along the bottom of the panel; they each require a 10 mm diameter mounting hole unless VC1 is a type having a three-hole fixing. This would then require a central mounting hole of about 8 mm in diameter and three smaller holes for the mounting screws. Usually 4BA mounting screws are used, which require mounting holes of about 4 mm in diameter. These screws fit through the mounting holes in the panel and into threaded holes in the front part of the capacitor. A simple way to find the positions for the smaller mounting holes in the panel is to make up a paper template by making a hole in a small piece of paper and then threading it on to the spindle of VC1. With the paper pressed up against the front of VC1 the positions of the mounting holes can be traced on to the paper to produce the template.

Note that VC1 can be any type having a maximum value of between about 200 pF and 300 pF, and inexpensive surplus types are perfectly suitable.

Circuit board

The space at the centre of the panel is occupied by a circuit board, which is constructed primarily from 0.1 matrix stripboard and either Soldercon pins or integrated circuit holders. Once the circuit demonstrator has been completed, this circuit board enables the various circuits to be quickly assembled as it uses solderless connection points, enabling the smaller components to be simply plugged into circuit and then unplugged again to clear the way for the next circuit.

Stripboard (normally sold under the proprietary name of Veroboard) consists of a sheet of insulating material such as SRBP (synthetic resin bonded paper) with a matrix of holes drilled in it. In this case the spacing is 0.1 inch between adjacent holes, but other matrix sizes are manufactured. Rows of holes are joined together on the underside of the board by continuous copper strips. See Figure 1.4. In normal use, circuit components are mounted on the top side of the board with their leadout wires protruding through holes in the board. The leadouts are cut to length and soldered to the copper strips. This gives the components a firm mounting, and the copper strips carry the interconnections between the components. By breaking strips at various points they can be made to carry several sets of connections if necessary.

In this unit sockets are soldered to the stripboard so that temporary circuits can be easily built and dismantled. The sockets are strips cut from dual-in-line ('DIL') integrated circuit holders. Each holder's two rows of sockets are carefully cut into singles using a hacksaw and, where necessary, cut to length. They are then filed to a neat finish. For example, the two rows of thirty sockets can be made from a 40 and

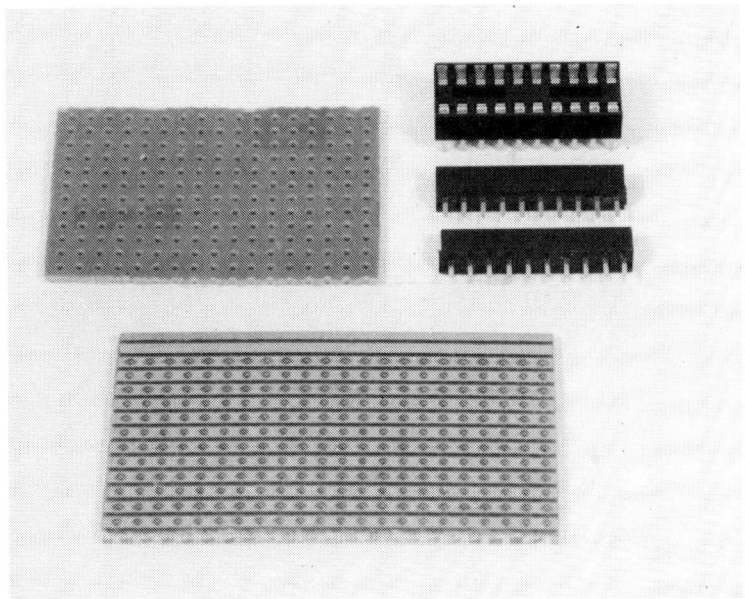

Figure 1.4 Stripboard (top and bottom views) plus a 'before' and 'after' 20 pin DIL IC holder

a 20 way integrated circuit holder, and two strips of five sockets can be made from a 14 pin holder. For the strips of three horizontal sockets it might actually be easier to use six vertical rows of eight sockets cut from 16 pin holders. Also, for the single sockets it might be easier to use 5 and 7 way strips with the pins of unused sockets trimmed off. Do not use the type of socket that has narrow entrance slits, as they will accept integrated circuit pins but not most component leads.

Details of the circuit board are provided in Figure 1.5. This requires a 0.1 inch matrix stripboard having 24 copper strips by 30 holes, and as the board is not sold in this size it must be cut from a large piece using a hacksaw. With 0.1-inch matrix board it is necessary to cut *through* rows of holes rather than *between* them; the rough edges that are produced can be filed down using a small file to give a neat finish. Next, make four mounting holes for 6BA or M3 clearance using a 3.3 mm drill, and break the copper strips at the 25 places marked with an 'X' in the diagram. A special tool for cutting the strips can be obtained, but it is also possible to use a hand-held twist drill about 4 mm or so in diameter, or even a sharp knife if due care is taken.

The strips of IC holder sockets can then be soldered into position. For this type of job a small 15—25 watt soldering iron is required, fitted

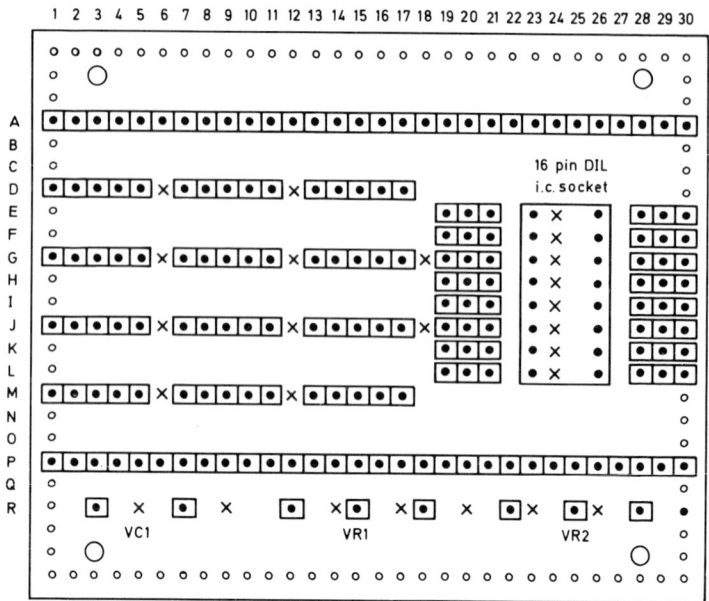

Figure 1.5 Details of circuit board

with a bit of 2—4 mm diameter. The solder should only be a resin-cored type for normal electronic and radio use (60/40 tin/lead type). This is generally available in 22 and 18 s.w.g. thicknesses, and either type is suitable.

In order to make a good soldered joint it is necessary to tin the iron with a small amount of solder, and to ensure that both the surfaces to be joined are clean and free from grease or corrosion. As the IC holder sockets are tin-plated and the copper strips on the board are coated with a protective layer of flux, there should be no problem here. In fact most modern components are designed so that they can be readily soldered into circuit.

For those new to soldering it would be advisable to practice soldering by connecting a few odds and ends of wire to a scrap piece of stripboard before attempting to connect the sockets into place. The bit of the iron and the solder should be applied to the joint simultaneously, rather than applying a lot of solder to the iron and then trying to transfer the solder to the joint. Using the first method the flux in the solder will aid the molten solder to flow over the surfaces to be joined, producing a good electrical and mechanical connection. With the second method the flux burns away before the solder is applied to the surfaces to be joined, making it hard to transfer the solder and probably producing a 'dry' joint.

Connecting the sockets is quite simple. With a strip of them slotted into position the board is laid strips uppermost on the work-top. This leaves both hands free, one to feed in the solder and the other to operate the soldering iron. Use plenty of solder so that it thoroughly covers the copper strips and the part of each socket that protrudes through the board. However, it is very easy to produce accidental short circuits between adjacent copper tracks, particularly on the right-hand section of this particular board, where there are a large number of connections in a confined area. Check carefully for such accidental short circuits and remove any that are spotted.

A 16 pin DIL integrated circuit holder is mounted on the right hand section of the board, and this is a whole socket that does not need any 'pruning'. The strips of sockets are quite densely packed around this, and it might be necessary to do some filing on these in order to fit them into the available space. Some makes of socket fit more easily than others, but it should not be too difficult to slightly file uncooperative components down to size.

Before finally mounting the circuit board on the panel it is necessary to drill three holes in the panel beneath the space occupied by the circuit board, as shown in Figure 1.3. It is also necessary to connect VR1, VR2 and VC1 to the eight single sockets at the front of the circuit board; these three components can then be connected into circuit via these sockets. The three holes in the main panel are needed to accommodate the connecting wires, which must obviously pass through the panel at some point; the exact diameter of these holes is unimportant. The eight connecting wires are soldered direct to the copper strips on the underside of the circuit board, with due care being taken not to dislodge the single sockets while doing this.

The finished circuit board is bolted to the main panel of the unit using short 6BA or M3 screws with the appropriate fixing nuts. The board itself can be used as a template to help mark the positions for the mounting holes in the panel. Shallow spacers are used over the mounting bolts to hold the board a little way clear of the main panel. If this is not done, the soldered joints on the underside of the board will probably cause the board to be distorted and possibly damaged as the fixing nuts are tightened.

Full wiring details of the unit are shown in Figure 1.6. A 1.5 kΩ resistor is mounted on the light-emitting diode D1, as shown in the diagram. It is important to note that D1 is not simply an ordinary light bulb, but a semiconductor device. In practice this means that it must be connected to the supply with the correct polarity or it will fail to light up. Also it must only be connected to the supply via the 1.5 kΩ current-limiting resistor or it may draw an excessive current and be

7

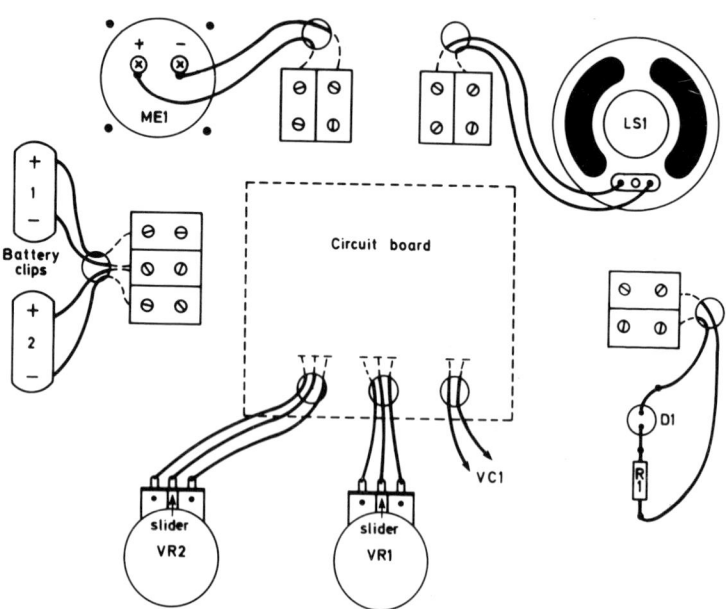

Figure 1.6 Wiring beneath top panel of circuit demonstrator unit

8

destroyed. The cathode normally has a slightly shorter leadout wire than the anode, but this is not invariably the case. With the cathode connected to the negative terminal of one battery, and the anode connected to the positive terminal of that battery via R1, D1 should light up. If it does not, try transposing the connections to it.

Finishing the construction

It is a good idea to fit the finished unit into some form of outer casing to provide it with a firm mounting and a neat finish (Figure 1.1). The casing for the prototype consists of four pieces of 95 X 25 mm timber. The front and rear sections are 266 mm long, the sides are 200 mm long, and the four pieces are held together by a total of eight large screws. Pieces of timber about 10 mm square are glued in place within the casing to act as supports to which the main panel is glued. One piece of timber on each side piece of the casing should be sufficient to give the main panel a firm mounting. The main panel is recessed about 10 mm into the outer casing. If desired, a base panel of thin plywood or good quality hardboard can be screwed to timber supports glued to the outer casing, and the outer casing can be painted. These are both far from essential though.

Control knobs must be fitted to VR1, VR2 and VC1, and if any of these are supplied with long spindles it will be necessary to trim them to a more suitable length. This is done using a hacksaw, with the spindle (*not* the body of the component) gripped in a vice. This should obviously be done before the control is finally installed in the circuit demonstrator unit.

In Figure 1.5 the strips and rows of vertical holes on the circuit board are identified by letters and numbers respectively. These correspond with references given on many of the circuits featured later in this book, and the purpose of the references is to show where on the board the various components are inserted to build up the circuit. For example, in most of the circuits battery 1 has its positive terminal connected to A-30 and its negative terminal connected to P-30. A-30 is obviously the extreme right-hand socket of the top row of sockets, and P-30 is the extreme right-hand socket of the long row of sockets towards the bottom of the circuit board (i.e. it is the socket that lies in copper strip 'P' and vertical row of holes number '30'.

In order to locate particular pins quickly it is helpful to mark on to the circuit board at least some of the reference letters and numbers. This can be done using rub-on transfers or some types of felt-tipped pen.

9

Components for circuit demonstrator unit

ME1	Moving-coil meter of 100 microamps sensitivity, with the front 42 mm square or 60 x 40 mm
D1	TIL220 (0.2 inch diameter) or TIL209 (0.125 inch diameter) red light-emitting diode with panel holder to suit
R1	1.5 kΩ (1500 ohms) $^1/_4$, $^1/_3$, or $^1/_2$ watt resistor, 5% tolerance
LS1	Miniature speaker of 50–75 mm diameter, having an impedance in the range of 40–80 ohms
VR1	470 kΩ linear carbon potentiometer
VR2	4.7 kΩ logarithmic carbon potentiometer
VC1	Variable capacitor having a maximum capacitance in the range 200 – 300 pF
By1	PP3 9 volt battery
By2	PP3 9 volt battery

Miscellaneous
Three control knobs
Two PP3-type battery connectors
12-way 5 A connector block
0.1-inch matrix stripboard having at least 24 copper strips by 30 holes
8 x 14 pin DIL integrated-circuit holders
4 x 16 pin DIL integrated-circuit holders
1 x 20 pin DIL integrated-circuit holder
1 x 40 pin DIL integrated-circuit holder
Wood, hardboard, etc. (see text)
Multistrand p.v.c.-insulated connecting wire, solder, mounting bolts, etc.

2
Passive components

This chapter is devoted to current flow and the main *passive* components (components that cannot provide amplification). It is important to have a basic understanding of the nature of current flow before one progresses to the function and operation of components and circuits. It should not simply be passed over as trivial and unimportant. Similarly, the role of the passive components is an essential one, and their function must be clearly understood before one can follow the action of even a simple practical circuit.

To illustrate the principles described, there are nine experiments for you to set up on your circuit demonstrator unit. The components needed are listed on page 37.

Free electrons

Although it is easy to show an effect of electric current flow, such as switching on an electric light, it is not possible to directly observe current flow. This is because it consists of minuscule particles, known as *electrons*, moving from one atom to another. Atoms are the basic units of every substance; each atom consists of a *nucleus* orbited by electrons, which compared with the nucleus have a negligible mass. The nucleus is formed from two types of particles known as protons and neutrons. The numbers of neutrons, protons and electrons that make up an atom are unique to the particular substance whose basic unit it is.

An electron is said to have a negative charge, and a proton a positive charge. The charge possessed by these particles can be regarded as a sort of basic quantity of electricity. Normally an atom has an equal number of protons and electrons and thus has no overall charge. However, certain materials (including common metals) have an electron in an outer orbit which is able to break away from the atom, and is known as a 'free electron'. It is this free electron that enables electric current flow to readily occur. Usually free electrons move randomly from one atom to another and there is no true current flow, but under certain circumstances they can be made to progress in the same direction and produce an electric current.

Conductors and insulators

As one would expect, a material that has an abundance of free electrons will more readily support a strong current flow than one that has few free electrons. Common metals such as copper and tin have plenty of free electrons, and can easily be made to produce a large current flow. Such materials are termed 'conductors'. Substances such as most plastics, rubber and glass have relatively few free electrons and require a considerable amount of persuasion before they will conduct a significant amount of current. These are known as 'insulators'.

The amount of current that flows is measured in units called 'amperes'. There is also a unit known as a 'coulomb' which takes time into account, and is actually a flow of one ampere for one second (or an equivalent amount such as 0.5 ampere for two seconds). The coulomb is little used in practical electronics, though, and the ampere is the commonly used unit of electric current flow.

In order to make the electrons flow in the same direction, rather than just moving at random, it is necessary to apply an electric force. A common source for such a force is an ordinary battery, which by chemical means generates an excess of electrons at one terminal (the negative or − terminal) and a shortage at the other terminal (the positive or + terminal). If a conductor is connected across the battery, free electrons will be attracted from the end of the conductor that is connected to the positive terminal, and will try to fill the deficit of electrons at this terminal. At the negative terminal the excess of electrons tends to spill out into the conductor. Since electrons tend to repel one another, this causes a general movement of electrons through the conductor from the negative end to the positive one. For as long as the battery remains charged, the free electrons flow out of the conductor into the positive terminal to try to fill the electron shortage, while at the other terminal the excess of electrons spills out into the conductor, forcing free electrons in the conductor along towards the positive terminal where they are needed to fill the deficit. This gives the required steady flow of electricity, until the battery's chemical energy becomes exhausted.

As will be apparent from the above, the electrons flow from negative to positive. However, current flow (or 'conventional current flow', as it is often termed) is always taken to be from positive to negative. This is because in the early days of electrical experimentation it was assumed that the current did actually flow from positive to negative, and a number of laws of electricity were based on this assumption. When it was discovered that electrons actually flow in the opposite direction, the idea of positive-to-negative current flow was well established and was not overthrown. Therefore we have a situation where *current* flow is said to be from positive to negative, and *electron* flow is said to be from negative to positive. The behaviour of most components and circuits can be explained using either system. In this book the convention of positive to negative current flow will be observed.

Voltage and resistance

The amount of current that flows through a circuit depends upon two factors; the power of the electrical force applied to the circuit, and how well or badly the circuit conducts electricity.

How powerfully the source of electricity tries to force the current through the circuit is measured in units called 'volts' (abbreviated to V). If one thinks of a wire carrying an electric current being analogous to a pipe carrying gas or water, the voltage is analogous to the pressure in the pipe. The higher the pressure, the greater the quantity of material (or electricity) that can be forced through a pipe (or wire) of a given diameter. It is important to differentiate clearly between voltage (the pressure or degree of force behind the source of electricity) and the current flow in amperes (the amount of electricity flowing through the circuit).

How well a material conducts electricity is termed 'conductance', but it is more normal to measure the ability of a conductor in terms of how much it *opposes* a current flow, rather than how easily it *permits* a current flow. The opposition to a current flow is termed 'resistance', and it is measured in 'ohms'.

One factor governing the resistance of a material is the quantity of material and how it is distributed. If we return to the analogy of water or gas being fed along a pipe, it requires less force to push a given amount of material through a short, thick pipe than it does to push it through a long, thin pipe constructed from the same amount of material. This is because there will be many more free electrons in a certain length of the thicker wire in relation to an identical length of the thinner wire. The larger number of free electrons permits an easier current flow. Also, the greater the length of material that the current must negotiate, the more times a free electron has to transfer from one atom to another, and the harder it is to force a given current through the material.

A second major factor governing the resistance of a material is simply how many free electrons there are within a given amount of material. The fewer free electrons there are to carry the current, the higher the resistance becomes. Another factor is the temperature of the substance. High temperatures tend to increase the level of random activity by free electrons, and this obviously hampers the well-ordered behaviour needed to give a low-resistance path. The resistance of a light bulb, for instance, rises considerably when it is switched on and becomes heated.

Ohm's law

Ohm's law gives an exact mathematical relationship between resistance, current and voltage, so if quantities for any two of these are known, a quantity for the third can be calculated. Ohm's law is:

$$\text{Current} = \frac{\text{voltage}}{\text{resistance}}, \quad \text{abbreviated to} \quad I = \frac{E}{R}$$

From the equation in this form it is possible to calculate the current that will flow if a known voltage is applied to a known resistance. For example, if 12 volts is applied to a 4 ohm resistance, the current flow in amperes (normally called 'amps' and abbreviated to A) is equal to 12/4 or 3 amps.

It is possible to carry out a few practical tests using the circuit demonstrator unit to prove Ohm's law. Start by setting up the simple circuit of Figure 2.1. As is normal practice in electronics, symbols to represent the components are used in the diagram, and the lines joining them represent the connecting wires. Meter ME1 and battery By1 are

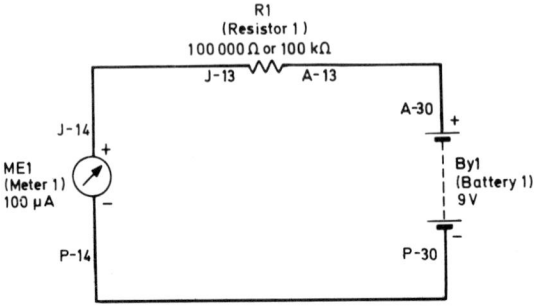

Figure 2.1 Simple circuit to demonstrate Ohm's law

Figure 2.2 Circuit shown in Figure 2.1 set up on the demonstrator unit

integral parts of the circuit demonstrator unit, but R1 is an additional component. Figure 2.2 shows the circuit working.

In this circuit we have a 9 volt battery feeding current through a 100 000 ohm resistor, with a current meter connected in series with the resistor to measure the current flow. In order to simplify matters we will assume that the meter itself has zero resistance. The meter does have a resistance of course, but this is likely to be only a few hundred ohms, which is so low in comparison with the 100 000 ohm resistor that it does not have any significant effect on the level of current flow.

From Ohm's law the current flowing in the circuit should be:

$$\frac{9 \text{ volts}}{100\ 000 \text{ ohms}} = 0.00009 \text{ amp}$$

The amp is rather a large unit, so in normal radio and electronics work it is more common to work in milliamps or microamps. A milliamp is one-thousandth of an amp and is usually abbreviated to mA. A microamp is equal to one-millionth of an amp and is normally abbreviated to μA. The meter used in the circuit demonstrator is calibrated from 0 to 100 μA, so the figure of 0.00009 amp must be multiplied by one million to give the expected meter reading in μA. As 0.00009 x 1 000 000 = 90, the expected reading is 90 μA.

Owing to the tolerances of the components used, it is unlikely that a reading of precisely 90 μA will be obtained in practice. For example, the 100 000 ohm resistor has a tolerance of ± 5%, which means that its actual value can be anything between 95 000 ohms and 105 000 ohms. The full scale deflection (f.s.d.) sensitivity of the meter and the battery voltage are also nominal figures. However, provided the battery is reasonably fresh the reading should be within a few µA of the calculated one.

As an exercise, replace R1 by various resistors with values between about 100 000 ohms and 1 000 000 ohms; calculate the *expected* current flow and compare it with the *actual* current flow.

The symbol used for the ohm is the Greek letter omega, Ω, and two further abbreviations are kΩ and MΩ. A kΩ or kilohm is equal to 1 000 ohms, and a MΩ or megohm is equal to 1 000 000 ohms. Thus a 330 000 ohm resistor would normally be said to have a value of 330 kΩ or even just 330k. A 6 800 000 ohm resistor would be said to have a value of 6.8 MΩ or even just 6.8M.

Alternative forms of Ohm's law

The form of Ohm's law used to find the resistance in the circuit when the voltage and current are known quantities is:

$$\text{Resistance} = \frac{\text{voltage}}{\text{current}}, \quad \text{abbreviated to} \quad R = \frac{E}{I}$$

Thus if a 10 volt source produces a current flow of 2 amps in a circuit, the resistance of the circuit is:

$$\frac{10 \text{ volts}}{2 \text{ amps}} = 5 \text{ ohms}$$

In the circuit of Figure 2.1 the battery voltage is 9 volts and the meter reads 90 µA. From this we can calculate that the value of R1 is:

$$\frac{9 \text{ volts}}{0.00009 \text{ amp}} = 100\ 000 \text{ ohms}$$

In other words the value should be 100 kΩ, which indeed it is.

The third form of Ohm's law is used to find the voltage applied to a circuit where the circuit resistance and level of current flow are the known quantities. The equation used is:

$$\text{Voltage} = \text{current} \times \text{resistance}, \quad \text{abbreviated to} \quad E = IR$$

Thus if a current of 3 amps flows though a 24 ohm resistor, the voltage applied to the resistor is

$$3 \text{ amps} \times 24 \text{ ohms} = 72 \text{ volts}$$

In the circuit of Figure 2.1 the current in the circuit is 90 µA or 0.00009 amperes and the resistance is 100 000 ohms. This gives a calculated voltage of 0.00009 amperes × 100 000 ohms = 9 volts, which is of course correct.

It is perhaps worth pointing out that Ohm's law is used a considerable amount when designing and testing electronic equipment, and it is well worthwhile committing it to memory. Provided you have a reasonably clear understanding of voltage, current and resistance this should not be difficult as Ohm's law should then seem perfectly logical and even obvious.

Colour coding of resistors

A practical problem that faces newcomers to electronics when dealing with resistors is the colour coding that is commonly used on these components to indicate their nominal values and tolerances. But before we look at colour coding it is worth mentioning that a few resistors simply have their value written on their bodies, although usually in a slightly abbreviated form. Examples of this method (which is sometimes used to show values on circuit diagrams as well) are given below. All the system does is to use a letter to indicate both the unit in use and the position of the decimal point, thus eliminating one digit in many cases.

33 or 33R = 33 ohms
2.2 or 2R2 = 2.2 ohms
68K = 68 kilohms
1M2 = 1.2 megohms

560 or 560R = 560 ohms
4K7 = 4.7 kilohms
220K = 220 kilohms
10M = 10 megohms

The tolerance is either written on as a percentage, or a code letter is used as follows: F = 1%, G = 2%, J = 5%, K = 10%, and M = 20%. Minor variations on this sytem are very occasionally encountered, but with a little common sense it should be easy to come to terms with them.

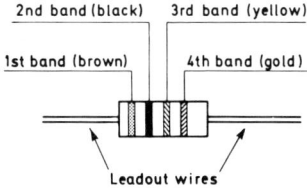

Figure 2.3 Value and tolerance of a resistor are usually indicated by four coloured bands (brown, black, yellow, gold = 100 kΩ, 5%)

By far the majority of resistors use a colour coding system where four coloured bands are marked around the body of the component, as shown in Figure 2.3. The first three bands indicate the component's value, the fourth signifies the tolerance. The first band is the one nearest to one end of the resistor's body, incidentally.

The first two bands give the initial two digits of the value, with colours corresponding to numbers as shown below.

17

Black	= 0	Green	= 5	
Brown	= 1	Blue	= 6	
Red	= 2	Violet	= 7	
Orange	= 3	Grey	= 8	
Yellow	= 4	White	= 9	

The third band is the multiplier, and indicates the number by which the first two digits must be multiplied to give the correct value. A list of colours and their corresponding multiplier values is given below.

Gold	= 0.1	Orange	= 1000	
Black	= 1	Yellow	= 10 000	
Brown	= 10	Green	= 100 000	
Red	= 100	Blue	= 1 000 000	

The fourth band indicates the tolerance, using the following code:

Brown	= 1%
Red	= 2%
Gold	= 5%
Silver	= 10%
None	= 20%

In the example in Figure 2.3 the first two bands are brown and black respectively, showing that the first two digits of the number are 1 and 0.

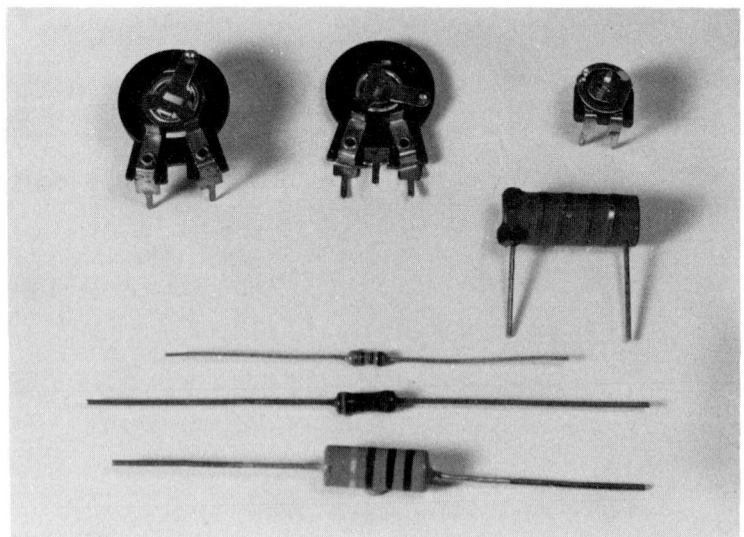

Figure 2.4 Selection of resistors. Top, left to right: horizontal preset, vertical preset, miniature horizontal preset (see later). Centre right: high-wattage resistor. Bottom: $\frac{1}{3}$ watt carbon film resistor, $\frac{1}{2}$ watt metal oxide resistor and 1 watt carbon composition resistor

The multiplier is yellow, indicating that the figure of 10 given by the first two bands must be multiplied by 10 000, giving a value of 100 000 ohms or 100 kΩ. The tolerance of the component is 5%, as the fourth band is a gold-coloured one. Thus the resistor has a marked value of 100 kΩ, and its actual value is guaranteed to be within the range 95 to 105 kΩ as it has a 5% tolerance.

To take another example, a resistor having a colour code of green, blue, gold, silver would have a value of 56 (green = 5, blue = 6) multiplied by 0.1 (gold multiplier band corresponds to 0.1) which equals 5.6 ohms. The silver fourth band indicates a tolerance of 10%.

Since a number of resistors are used in virtually every piece of electronic equipment it is a good idea to become familiar with the resistor colour code as quickly as possible. The code may seem an unnecessary complication, but it does have advantages, one of which is that this form of marking is unlikely to wear off, or partially wear off so that the value of the component is misread. With experience one learns to recognise the various values on sight, and this brings the further advantage of being able to pick out a component of the required value quickly from an assortment of resistors. Some are shown in Figure 2.4.

Series resistance

Returning now to more theoretical matters concerning resistors, consider the circuit of Figure 2.5. This is similar to the circuit of Figure

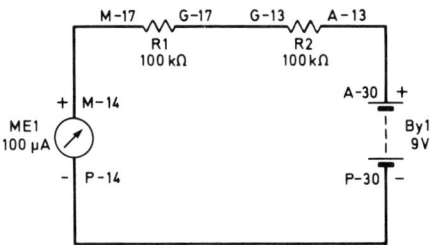

Figure 2.5 Simple series-resistance circuit

2.1, but an additional resistor has been added into the chain of components in such a way that the current has to force its way first through R1, and then through R2. This is known as a 'series' circuit. What current flows in the circuit, and what is the combined resistance of the two resistors?

If the circuit is constructed on the circuit demonstrator unit it should be found that the current flowing in the circuit is about 45 μA. From Ohm's law the total circuit resistance can be calculated, and is:

$$R = \frac{E}{I} = \frac{9 \text{ volts}}{0.000045 \text{ amp}} = 200\,000 \text{ Ω or } 200 \text{ kΩ}$$

19

In other words the two 100 kΩ resistances have been added together to effectively form a 200 kΩ component. Since the current has to force its way through first one resistor and then the other, this combining of the two values to produce an effective value equal to their sum is really what one would expect. In fact the total resistance of a series resistor circuit is *always* equal to the sum of the resistor values, regardless of the number of resistors present in the circuit.

This can be expressed by the equation:

$$R_t = R_1 + R_2 + R_3 + R_4 \ldots$$

where R_t is the total circuit resistance and R_1, R_2, R_3, etc. are the individual resistance values in the circuit. Thus, for example, a 5.6 kΩ resistor in series with a 10 kΩ resistor and a 22 kΩ component will give a total resistance of 5.6 + 10 + 22 = 37.6 kΩ.

Parallel resistance

Figure 2.6 shows an alternative way of combining two (or more) resistors: this time they are simply connected across one another, or what is termed 'in parallel'. What current flows in this case, and what is the combined resistance of the two circuits?

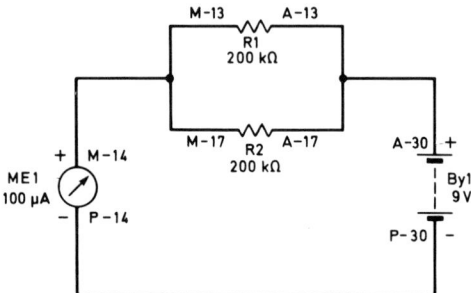

Figure 2.6 Simple parallel-resistance circuit

If the circuit is constructed on the circuit demonstrator it should be found that a current of about 90 µA is indicated by ME1. This is really what one would expect as 9 volts is applied to R1, and from Ohm's law it is apparent that it will therefore pass 45 µA:

$$I = \frac{E}{R} = \frac{9 \text{ volts}}{200\,000 \text{ ohms}} = 0.000045 \text{ amp or } 45 \text{ µA}$$

R2 has the same value and applied voltage, and therefore passes the same amount of current. With 45 µA passing through both R1 and R2 there is obviously a total current of 90 µA flowing through the circuit.

The effective combined value of the two resistors is obviously 100 kΩ, as we saw in an earlier example that a current flow of 90 μA results from 9 volts being applied to a 100 kΩ resistor:

$$R = \frac{E}{I} = \frac{9 \text{ volts}}{0.00009 \text{ amp}} = 100\,000 \text{ Ω or } 100 \text{ kΩ}$$

One would expect the combined resistance of this configuration to be less than the value of any resistor in the circuit, since there are additional current paths to that lowest-value component. The equation for calculating the combined resistance of a parallel-resistance circuit is:

$$R_t = \frac{1}{\frac{1}{R_1} + \frac{1}{R_2} + \frac{1}{R_3} + \ldots}$$

If we take a simple example, say two eight-ohm resistors and one four-ohm resistor in parallel, this gives:

$$\frac{1}{\frac{1}{8} + \frac{1}{8} + \frac{1}{4}} = \frac{1}{0.125 + 0.125 + 0.25} = \frac{1}{0.5} = 2 \text{ Ω}$$

Where there are only two resistances in parallel, it is possible to use this alternative equation:

$$R_t = \frac{R_1 \times R_2}{R_1 + R_2}$$

For example, the effective resistance of a 100 ohm resistor and a 400 ohm resistor connected in parallel would be:

$$\frac{100 \times 400}{100 + 400} = \frac{40\,000}{500} = 80 \text{ Ω}$$

Voltmeter

In the next experiment we require a meter to measure voltage rather than current, and this is achieved by simply adding a 100 kΩ resistor in series with the meter. This combination then acts as a 0–10 volt voltmeter.

This can be proved with the aid of Ohm's law. The resistance of the meter is negligible in relation to that of the 100 kΩ resistor, and so from Ohm's law we know that the voltage needed to force 100 μA through the circuit and give full scale deflection of the meter is:

$$E = IR = 100\,000 \times 0.0001 = 10 \text{ volts}$$

Thus the circuit does indeed have the required f.s.d. value of 10 volts.

A lower voltage would produce a proportionately lower meter reading. For example, two volts would produce a reading of 20 µA:

$$I = \frac{E}{R} = \frac{2}{100\,000} = 0.00002 \text{ amp or } 20 \text{ µA}$$

In other words, dividing the scale reading by ten gives a reading in volts.

Kirchhoff's voltage law

In the circuit of Figure 2.7 there is obviously a total series resistance of 9 kΩ (3.3 + 1.8 + 1.2 + 2.7 = 9), and from this the current flow can be calculated as 1 mA:

$$I = \frac{E}{R} = \frac{9}{9000} = 0.001 \text{ amp or } 1 \text{ mA}$$

The voltage developed across each resistor can then be calculated; the voltage across R1, for example, will be:

$$E = IR = 0.001 \times 3300 = 3.3 \text{ volts}$$

Calculating the voltages developed across R2 to R4 should give answers of 1.8 volts, 1.2 volts and 2.7 volts respectively. The voltmeter made by

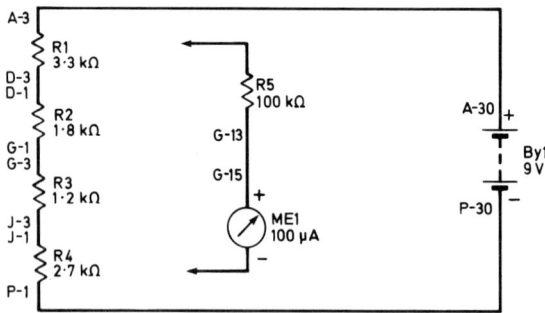

Figure 2.7 Circuit to demonstrate Kirchhoff's law; R5 and ME1 form a 0–10 V voltmeter

adding a 100 kΩ resistor in series with ME1 can also be used to measure these voltages and confirm the calculations. Make sure that the meter is connected to the circuit with the correct polarity (the negative terminal always connects to the lower leadout wire of each resistor).

The total of the voltages developed across each of these resistors comes to 9 volts, and is equal to the voltage fed into the system. However many resistors are added in series, and whatever their values, the total of the voltages developed across them will always be equal to the voltage applied to the circuit, and this is known as Kirchhoff's voltage law or simply as Kirchhoff's law. This might seem obvious, but it is an important principle which you should understand before considering the operation of practical circuits.

The voltage developed across two or more resistors in a circuit of this type is of course equal to the sum of the voltages across the individual resistances. For example, from the negative supply rail to the junction of R2 and R3 there is 3.9 volts, and from the negative rail to the junction of R1 and R2 there is 5.7 volts. Again this may seem an obvious point, but it is an important one. This type of circuit is known as a *voltage divider* or a *potential divider*, and can be used when a supply voltage lower than the main supply voltage is required.

Variable resistors

A variable resistor has a 'track' in the shape of a broken circle, made from a deposit of carbon or from resistance wire wound on to a former constructed from an insulating material. An electrical contact is connected to each end of the track, and there is a fixed level of resistance between these two points. There is a third contact, which can be moved along the track by means of a rotatable spindle. The resistance between this contact and one of the end contacts can be varied by altering the position of the moving contact, since the greater the amount of track between the two points, the greater the resistance between them. Thus the resistance of the component can be varied from zero to the full track resistance.

By using all three contacts it is possible to use a variable resistor in a potential-divider circuit to give a variable voltage, as shown in the circuit of Figure 2.8 (note the circuit symbol for VR1). Here the voltage reading on ME1 can be continuously varied from zero when it is

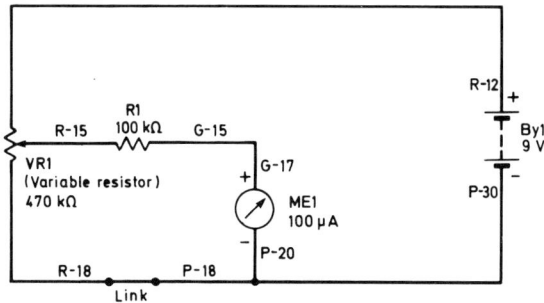

Figure 2.8 Use of a potentiometer to give continuously variable voltage at its slider

23

in a fully anticlockwise position, to the full supply voltage of 9 volts when it is adjusted to a fully clockwise position. When all three connections of a variable resistor are used it is more correctly called a *potentiometer*, and this is in fact the name generally used for these components.

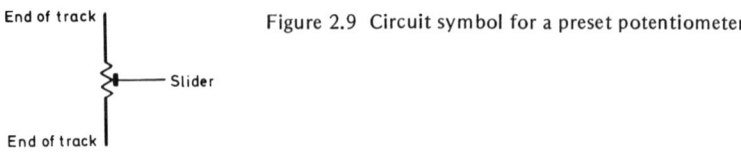

Figure 2.9 Circuit symbol for a preset potentiometer

There is a type of potentiometer that has an open 'skeleton' type construction, and is adjusted by means of a screwdriver. These are preset potentiometers or just 'presets', and have the circuit symbol shown in Figure 2.9. Examination of one of these components will

Figure 2.10 **Preset potentiometer:** this view shows the carbon track and the wiper contact

clearly show the general construction and operation of a potentiometer. See Figures 2.4 and 2.10.

Power

When current flows in a resistor it dissipates energy in the form of heat production. The power dissipated is given by the equation:

Power = voltage x current,   abbreviated to $P = EI$

This gives the answer in units known as 'watts'. In the same way that a milliamp is a thousandth of an amp, a milliwatt is a thousandth of a watt.

If we take a simple mathematical example, a resistor having a value of 10 ohms would pass a current of 0.5 amp if it were subjected to a voltage of 5 volts. The power it would dissipate is equal to 0.5 amp x 5 volts, i.e. 2.5 watts.

Resistors have a 'power rating', which should not be exceeded as this could easily result in the component being overheated and destroyed. In the circuits provided in this book the resistors only have to dissipate minute power levels; the total power dissipated in R1 to R4 of Figure 2.7, for instance, is only 9 mW (9 volts x 0.001 amp = 0.009 watt or 9 mW). Owing to these low dissipation levels it is perfectly acceptable to use miniature resistors having power ratings of $1/8$, $1/4$ or $1/3$ watt. In fact these are preferable to larger types, which would not fit on to the circuit demonstrator unit quite as easily.

Capacitors

Capacitors occur in electronic circuits almost as frequently as resistors, but the functions of the two are completely different. A capacitor consists of two plates made from a material that is a good conductor, separated by a thin layer of a good insulating material, as shown in Figure 2.11. On the face of it a capacitor is of little use since it would ideally pass no current whatever, but it does have certain very useful applications, as we shall see later.

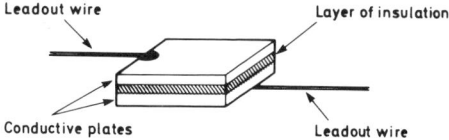

Figure 2.11 General method of construction of a capacitor

The main property of a capacitor is its ability to *store* an electrical charge, and this can be demonstrated using the circuit of Figure 2.12. The circuit symbol for a capacitor is two parallel lines, which represent the two conductive plates; the gap between the lines represents the layer of insulation. This symbol is shown in Figure 2.13(a). The capacitor used in the circuit of Figure 2.12 is a special type known as an *electrolytic*, and this uses a substance known as an electrolyte to provide the insulating layer between the conductive plates. It enables a physically small component of high value to be produced, but this type of capacitor will only work if it is fed with a supply of the correct polarity. As will be apparent from Figure 2.12, the circuit symbol for

25

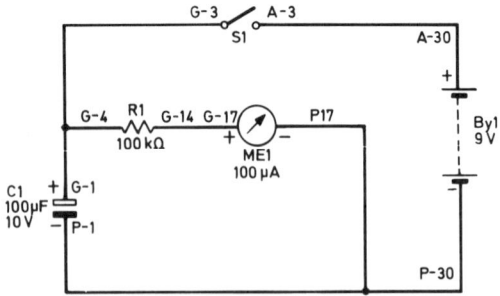

Figure 2.12 Simple circuit to demonstrate charge storage in a capacitor

an electrolytic capacitor is slightly different to that for an ordinary type, as it must indicate the correct polarity. The plate that is shown as a solid line is the negative terminal, and the one that is shown in outline is the positive terminal. In practice '+' and '−' signs are usually marked

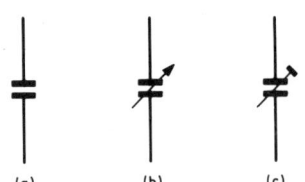

Figure 2.13 Circuit symbols for (a) ordinary (non-electrolytic) capacitor, (b) variable capacitor, (c) preset capacitor

on the component to indicate its polarity, and it is also common for the positive terminal to be indicated by an indentation around the body of the component at the appropriate end of the component. Typical capacitors are shown in Figure 2.14.

In the circuit of Figure 2.12, if you set this up on the circuit demonstrator unit, S1 does not have to be a proper switch; it can be a link wire connected in position to simulate the closed switch, and removed to simulate the switch in the open position. With this wire link placed in position, the battery is connected across the capacitor. This causes free electrons to be attracted from the positive terminal of C1 into the positive side of the battery, and a similar number of free electrons to spill out from the negative battery terminal into the negative plate of C1. As this happens virtually instantly, ME1 registers the supply potential of 9 volts and does not indicate that anything has happened.

If S1 is then opened (or the link wire simulating S1 is removed), the charge that C1 has received from the battery, and has stored, will leak away through the voltmeter circuit, because the excess of free electrons on the negative plate of C1 is attracted to the positive plate with its deficit of free electrons. This leakage results in a reading being obtained from the meter, even though the battery has been disconnected.

26

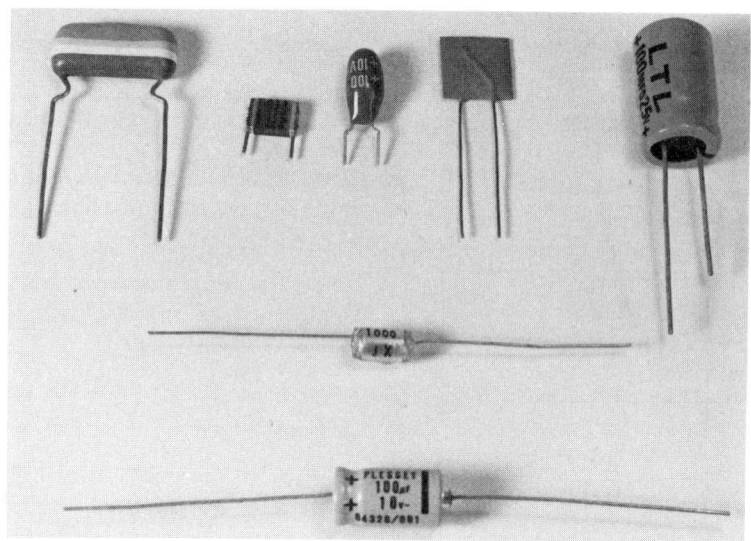

Figure 2.14 Selection of capacitors. Top, left to right: two plastic foil radial-lead (printed circuit mounting) types, a tantalum bead type (high-quality miniature electrolytic), a ceramic-plate type, and a radial-lead electrolytic. Below: a polystyrene and an electrolytic type

Initially the reading is 9 volts, but this gradually diminishes as the charge on C1 dies away. At first the meter needle moves fairly quickly towards zero, but as time passes it moves more and more slowly. This happens because the initial 9 volt charge forces 90 μA through the voltmeter circuit. When the charge on C1 has fallen to (say) 2 volts, only 20 μA is forced through the voltmeter circuit. This decreasing level of discharge current causes the drop in voltage across C1 to slow up gradually. It discharges 'exponentially'.

If the voltmeter circuit is removed, C1 will theoretically remain charged indefinitely, although in practice the insulation between the two plates of the capacitor is not perfect, and the component will gradually discharge through it. The layer of insulation is termed the *dielectric*, and the current that flows through the dielectric is the *leakage*. Incidentally, like resistors, capacitors are available in variable and preset versions (although only in low values), and these have the circuit symbols of Figure 2.13(b) and (c) respectively.

Charge time

In the circuit of Figure 2.12 the capacitor charges almost instantly when S1 is closed, because the battery is able to supply the full charge within a very short space of time. If the capacitor were to be charged via a circuit having a fairly high resistance, this would severely limit the available charge current and slow up the charging process, in much the same way that C1 slowly discharged through the voltmeter circuit when S1 was opened in the Figure 2.12 experiment.

This effect can be demonstrated by the test circuit shown in Figure 2.15. When S1 is closed (as before, this can simply consist of a link wire

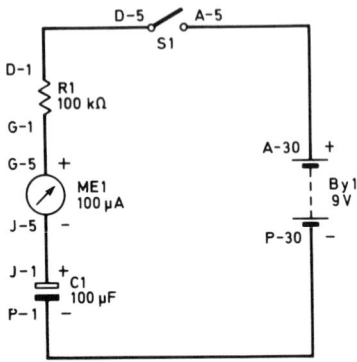

Figure 2.15 Circuit to demonstrate the charge time of a capacitor

if the circuit is constructed on the circuit demonstrator unit), C1 begins to charge up via the relatively high resistance of the voltmeter circuit formed by R1 and ME1. Initially there will be no charge voltage on C1, so the full 9 volt supply appears across the voltmeter and ME1 registers a reading of 9 volts. ME1 is also registering the charge current of C1, which is obviously 90 µA.

As C1 charges up, the voltage across this component rises; the potential across the voltmeter therefore decreases, as its share of the 9 volt supply must diminish. This will be indicated by a falling reading on ME1. As the voltage across the voltmeter decreases, the current forced through this circuit also decreases. This is also the charge current for C1, of course, and so the charging of C1 gradually slows down. This effect is demonstrated by the falling charge current indicated on ME1, and the slowing of the speed at which ME1's pointer moves towards the zero mark. It is charging in an exponential manner.

Capacitance value

Capacitance is measured in a basic unit called the 'farad', and is defined as the charge required to bring the conductive plates to a given voltage difference. One farad equals a charge of one coulomb to give a voltage difference across the capacitor of one volt. In other words, a current of one amp must be fed into a one-farad capacitor for one second in order to charge it to one volt. An equivalent charge such as 0.25 amp for 4 seconds would have the same effect. To charge a one-farad capacitor to a higher voltage takes a proportionately higher charge, 3 coulombs to give 3 volts for example.

The farad is far too large a unit for use in general electronic work, and the following fractional units are normally used:

- microfarad, normally abbreviated to μF, and equal to one-millionth of a farad;
- nanofarad, normally abbreviated to nF, and equal to one-thousandth of one-millionth of a farad;
- picofarad, normally abbreviated to pF, and equal to one-millionth of one-millionth of a farad.

Thus there are 1000 pF in 1 nF, 1000 nF in 1 μF, and 1 000 000 μF in one farad.

Time constants

The time taken for a capacitor to charge via a resistor to 63% of the supply voltage, or to discharge via a resistor to 37% of the initial charge voltage, is CR seconds — its *time constant*. Rather than have C in farads and R in ohms, in most practical calculations it is more convenient to use μF and MΩ. As an example, if a 10 μF capacitor is charged via a 2.2 MΩ resistor, the voltage on the capacitor will equal 63% of the supply potential after 10 x 2.2 = 22 seconds. In the circuit of Figure 2.12 a 100 μF capacitor is being discharged through a 100 kΩ resistor, and so 37% of the initial charge voltage should be reached after 100 x 0.1 = 10 seconds.

In theory a capacitor never achieves full charge if it is charged via a resistor, or complete discharge if it is discharged by a resistor, but more than 99% charge or discharge will be achieved in $5CR$ seconds.

Series and parallel capacitors

If two or more capacitors are connected in parallel their combined value is equal to the sum of their values, since in order to charge or discharge one capacitor it is necessary to simultaneously charge or discharge all the others. The equation for parallel-connected capacitors is:

$$C_t = C_1 + C_2 + C_3 + \ldots$$

The equation for series-connected capacitors is:

$$C_{\text{effective}} = \frac{1}{\frac{1}{C_1} + \frac{1}{C_2} + \frac{1}{C_3} + \ldots}$$

Note that the parallel-capacitance and series-resistance equations are similar, as are the series-capacitance and parallel-resistance ones.

A.C. coupling

As stated earlier, no current actually passes through a capacitor, except for a minute leakage current that occurs with practical components. However, a capacitor can be used to effectively pass varying d.c. (direct current) or a.c. (alternating current), while blocking straightforward d.c. signals. Before demonstrating this process we must understand clearly what is meant by d.c., varying d.c., and a.c.

A battery is a good example of a d.c. source. If a circuit is connected across a battery, there is a (conventional) current flow from the positive to the negative terminal, and the battery always produces a current flow in this same direction. The voltage provided by a battery is reasonably constant, and so it provides what could be termed a *fixed* or *steady d.c.*

If a potentiometer is connected across a battery, and a circuit is connected between the slider of the potentiometer and one of its track connections (as in Figure 2.8 earlier, for example), the circuit is still provided with a d.c. supply since the current flow is always in the same direction. Adjusting the setting of the potentiometer backwards and forwards gives a varying voltage, but does not affect the polarity of the signal. This is a *varying d.c.* signal.

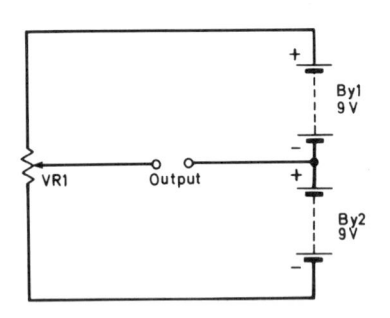

Figure 2.16 Simple bridge circuit for generating an a.c signal

The situation is somewhat different in the circuit of Figure 2.16, where the battery is connected in series with a second battery so as to give a total supply potential of 18 volts. The potentiometer is wired across both batteries, and the output is obtained from between its slider and the junction of the two batteries. The voltage at the output is equal to the difference between the voltages at these two points. This type of circuit occurs quite commonly in electronics, incidentally, and is known as a 'bridge'.

There is of course +9 volts at the junction of the two batteries relative to the lower supply rail. With the slider of VR1 half-way up its track there will be +9 volts at the slider also. This gives zero voltage across the output terminals. Taking the VR1 slider further up its track causes the slider voltage to go above 9 volts, and with the slider right at the top of its track it is actually 9 volts positive with respect to the other output. Intermediate settings give intermediate levels of positive output voltage. If the VR1 slider is taken *below* the centre track position it falls below 9 volts with respect to the lower supply rail, and negative relative to the other output of the bridge; it can achieve a maximum of −9 volts.

Therefore with this circuit it is possible to obtain an output that varies in both amplitude (voltage) and polarity, simply by repeatedly adjusting VR1 slider backwards and forwards, with the movement being centred on the centre of its track. A signal that varies in both amplitude and polarity is called an *alternating current*, commonly abbreviated to a.c. One complete 'cycle' of an a.c. signal consists of the output rising from zero to its peak voltage, crossing back through the zero point and continuing to the peak level of the opposite polarity, and then returning to zero. The *frequency* of an a.c. signal is the number of cycles in a one-second period, and the unit in which frequency is measured is the 'hertz'. Thus if 2000 cycles occur within a one-second period the signal has a frequency of 2000 hertz, or two kilohertz. This would normally be abbreviated to 2000 Hz or 2 kHz (1 kHz equals 1000 Hz). In radio work frequencies of many million hertz are often encountered, and frequencies of this order are usually expressed in megahertz (1 MHz equals one million hertz). The prefixes 'kilo' and

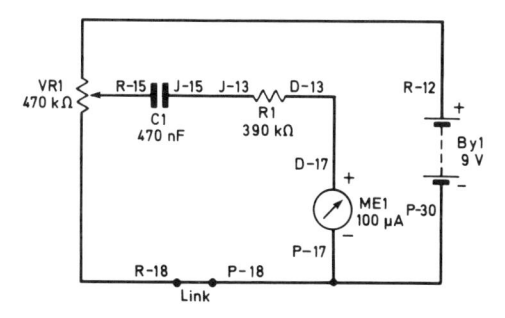

Figure 2.17 Circuit to demonstrate the a.c. coupling effect of a capacitor

'mega', also used earlier in the section dealing with resistance values, can in fact be used ahead of any unit to indicate multiples of 1000 and 1 000 000 respectively.

We can now consider the circuit of Figure 2.17, which illustrates how a varying d.c. signal or an a.c. signal can be effectively coupled through a capacitor. Note that no current actually flows through the capacitor; it simply gives an effect that in practice is much the same as if the signal were actually passing through the component.

If VR1's slider is initially set at the bottom of its track, nothing will happen when power is first applied to the circuit, apart from a small current flow through VR1. Adjusting the slider of VR1 to the other end of its track causes a voltage to be applied to C1 from the battery via R1 and ME1, and a charge current flows into C1. This is indicated by a small positive deflection of the meter. Moving VR1's slider back to the bottom end of the track removes the battery voltage from C1, and provides it with a discharge path through R1 and ME1. The end of C1 that connects to R1 has a negative charge, but connects by way of R1 to the meter's positive terminal. The discharge current of C1 therefore produces a small reverse deflection of the meter. If you set up the circuit on the demonstrator unit, it is advisable to temporarily adjust the zero set screw of the meter (the screwhead on the front of the meter) to give a quiescent reading of one or two scale divisions. This reverse deflection will then be more clearly seen.

If the slider of VR1 is continuously moved from one end of its track to the other, C1 is repeatedly charged and discharged, causing deflections of the meter. If it is moved backwards and forwards over only a small part of the track the voltage fed to C1 will still vary to some extent, and the consequent charge and discharge currents of C1 will still produce deflections of the meter. The currents and the deflections will both be smaller, though. If adjustment of VR1 ceases, C1 quickly assumes a charge voltage equal to that supplied by VR1, and no further current is coupled to R1 and ME1. Thus a capacitor can couple a varying signal from one part of a circuit to another, and will block d.c. levels.

Reactance

The effective resistance of a capacitor to a signal is termed *reactance*, and is affected by frequency as well as capacitance value.

This can easily be demonstrated by experimenting with the circuit of Figure 2.17. Adjusting VR1 quickly to give a relatively high frequency gives stronger meter deflections than if it is adjusted slowly. With slow changes in the slider voltage of VR1, C1 is able to maintain a charge voltage virtually equal to this slider voltage. This gives little voltage across R1 and ME1, with a low resultant charge current. Faster voltage changes produce a greater voltage difference and charge current, since the relatively slow time constant of R1 plus ME1 and C1 means that the voltage across C1 lags well behind that at the VR1 slider. Increasing the value of C1 increases the time constant and has the same effect. Reactance therefore *decreases* with both increasing capacitance and increasing frequency.

Reactance is given by the equation:

$$X_c = \frac{1\,000\,000}{2\pi fC}$$

where X_c is capacitive reactance in ohms, f is the frequency in hertz, and C is the capacitance value in microfarads. Thus, for example, the 470 nF (0.47 µF) capacitor of Figure 2.17 has a reactance of just under 34 kΩ at a frequency of 10 Hz. In practice it is quite common to use charts or tables to find reactance values, if they do not need to be known with a very high degree of accuracy.

Inductors

An inductor could be regarded as having properties that are the exact opposite of those possessed by a capacitor. It has a theoretical resistance of zero, and a level of reactance that *increases* with increasing inductance value and/or frequency. This is termed *inductive reactance* to differentiate it from capacitive reactance, which decreases with increasing capacitance value and/or frequency, as mentioned above.

An inductor can consist simply of a coil of wire, but the coil is usually wound on a core made from a material that gives a high inductance value for a given number of turns. This type of component functions in the way it does because of two electromagnetic phenomena. The first is that a magnetic force is produced around an inductor carrying an electric current. Varying the level of the current varies the strength of the magnetic field. It is this effect that is used in the operation of moving-coil meters and loudspeakers such as those used on the circuit demonstrator unit. It is also used in electric motors and many other items of electrical and electronic equipment.

The second (in fact, reverse) effect occurs when a coil is placed in a varying magnetic field; a varying voltage is produced. This effect is used in dynamos, moving-coil microphones, and many other items of equipment.

When an a.c. signal is applied to an inductor it generates an alternating magnetic field, and this field generates a second a.c. signal in the coil. The generated signal is of opposite polarity to the input signal and therefore opposes it. This obviously hinders the flow of the input signal through the coil.

The basic unit of inductance is the henry, and an inductor has a value of 1 henry when an applied potential of 1 volt gives a change in current of 1 amp per second. The henry is an extremely large unit, so the following sub-units are more commonly used:

- millihenry, abbreviated to mH and equal to one-thousandth of a henry;
- microhenry, abbreviated to µH and equal to one-millionth of a henry.

The reactance of an inductor rises with increased frequency because, no matter how fast the input voltage rises and falls, the rate at which the current flow varies is determined by the input voltage and inductance value. This is obviously of little hindrance to a very low-frequency signal, which would not produce very rapid variations in current flow anyway, but will severely limit the current flow produced by a high-frequency input signal. Such a signal quickly reaches its peak value, and then drops back to zero before the current flow has a chance to reach significant proportions. Increased inductance value gives a longer time constant, and thus also results in higher levels of reactance. The reactance of an inductor is given by the equation:

$$X_L = 2\pi f L$$

where X_L is the inductive reactance in ohms, f is the frequency in hertz, and L is the inductance value in henries.

A term often encountered in electronics is *impedance*. It is used where a circuit has some combination of resistance, capacitive reactance and inductive reactance. Impedance is defined as peak voltage divided by peak current.

Transformers

If a varying current is passed through an inductor a magnetic field of varying strength is produced. If a second inductor is placed within this magnetic field an alternating voltage is produced across the second inductor. This basic arrangement forms what is known as a *transformer*.

The transformer is of great importance, because it enables a low input voltage to give a much higher output voltage. The theoretical output voltage (V_{out}) of a transformer is given by the equation:

$$V_{out} = \frac{V_{in} \times \text{secondary turns}}{\text{primary turns}}$$

where V_{in} is the input voltage, secondary turns is the number of turns on the output winding (secondary winding), and primary turns is the number of turns on the input winding (primary winding). In other words, a transformer having (say) ten times as many turns in the secondary winding as in its primary winding will give an output voltage ten times higher than that applied at its input. Note that it is the *ratio* of primary turns to secondary turns that determines the step-up ratio of the transformer; the actual numbers of turns used to give the required ratio are not relevant in theory.

It must be pointed out that a transformer cannot provide *power* gain; an output voltage higher than the input voltage can only be obtained if the input current is higher than the output current by a factor at least equal to the step-up ratio of the transformer. Thus if 10 volts at 100 mA is drawn from the output of a transformer having a 10:1 turns ratio, an input of at least 1 volt at 1 amp is required. Practical transformers usually have an efficiency of about 60 – 70%, so they actually give a significant power loss.

Many transformers have fewer turns on the secondary winding than on the primary, and therefore give a *reduction* in voltage. This technique can be used to step-down the 240-volt mains supply to a low voltage, to supply equipment such as transistor radios, which require only about 9 volts. A voltage divider made from a couple of resistors could be used to do this, but the current drawn from the mains would be equal to that required by the equipment. When a transformer provides a voltage step-down it can give a similar step-up in current, and is therefore usually much more efficient than a voltage divider. In the particular application mentioned above, a transformer also has the advantage of isolating the equipment from the mains since there is no direct connection through the transformer, and this is good from the safety aspect.

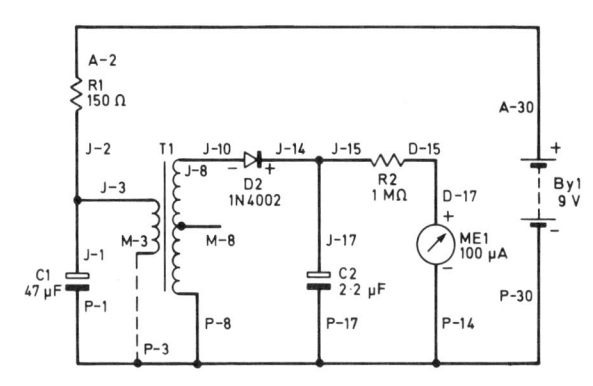

Figure 2.18 Circuit to demonstrate the function of a transformer

The circuit of Figure 2.18 can be used to demonstrate the voltage step-up ability of a transformer. R2 and ME1 form a voltmeter circuit, and from Ohm's law it will be seen that this circuit has an f.s.d. sensitivity of 100 volts ($E = IR$ = 0.0001 x 1 000 000 = 100 V). D2 is a diode, and together with C2 it converts the a.c. signal it receives from transformer T1 to a reasonably steady d.c. signal. (This is dealt with in Chapter 3 and will not be considered in detail here.)

C1 quickly charges up to the full supply-rail potential when power is applied to the circuit, and it will rapidly be largely discharged through the smaller winding of T1 if a shorting wire is temporarily connected between points M-3 and P-3. The low voltage applied to the small winding of T1 (which acts as the primary in this case) causes a higher

35

voltage to be induced in the larger secondary winding, and gives a voltage reading from the meter. ME1 should actually jump to a reading of about 15 volts or so, and then rapidly drop back to zero. In order to obtain a more accurate and sustained reading it is necessary to repeatedly connect and disconnect the wire link so that a fairly rapid succession of pulses is fed to T1 as C1 is first charged via R1 and then discharged into T1. This should give a reading of about 45 volts on the meter, indicating a step-up of about 5:1.

If the negative terminal of D2 is moved to terminal M-8, repeating the experiment will give a voltage reading of only about half the previous level. This is because D2 is taken to a connection at the centre of the large winding, thus effectively halving the number of secondary turns. Incidentally, a connection made to a point other than one of the ends of a coil is called a 'tapping' or a 'tap'.

The reason for feeding the primary winding of T1 from the main supply via R1 and C1, rather than direct, is because a current-limiting resistor must be added in series with T1 to prevent it from drawing an

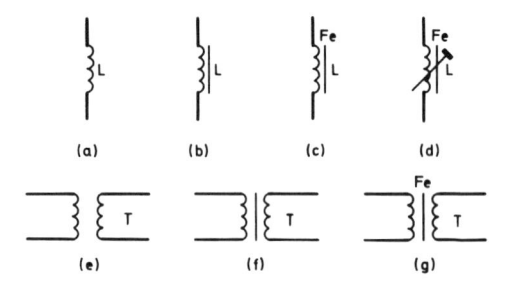

Figure 2.19 Circuit symbols: (a) air-cored inductor, (b) iron-cored inductor, (c) ferrite-cored inductor, (d) variable inductor (adjustable ferrite core), (e) air-cored transformer, (f) iron-cored transformer, (g) ferrite-cored transformer

Figure 2.20 Left: a mains transformer. Right: a small audio transformer

excessive current while the link wire is in place. C1 is then needed to provide a large but brief discharge current into T1, as the current available through R1 is not sufficient to give good results.

Finally, the circuit symbols for inductors and transformers vary according to the core material and type. The various symbols in common use are shown in Figure 2.19. Typical transformers are shown in Figure 2.20.

Components used in Chapter 2

Resistors (all miniature $^1/_8$, $^1/_4$ or $^1/_3$ watt, 5% tolerance)
150 Ω
1.2 kΩ
1.8 kΩ
2.7 kΩ
3.3 kΩ
100 kΩ (2 required)
200 kΩ (2 required)
390 kΩ
1 MΩ

Capacitors
470 nF plastic foil (polyester, etc.)
2.2 μF 10 V electrolytic
47 μF 10 V electrolytic
100 μF 10V electrolytic
Note that electrolytics having *higher* maximum working voltage than the specified figure are perfectly suitable, but types having *lower* voltage ratings are of course unsuitable.

Semiconductor
Diode type 1N4002

Transformer
Transistor output transformer type LT700

3
Semiconductor devices

The way in which semiconductor devices function is rather complicated, and it is probably better for the beginner to concentrate initially on what these components do and how they are used in practical circuits, rather than on their detailed internal operation. Therefore this chapter is devoted to the characteristics of the main semiconductor devices and their applications.

There are twelve more experiments to try on the circuit demonstrator unit. The additional components needed are listed on pages 66–67.

Diodes

Diodes have the characteristic of allowing a current flow in one direction only. They are also known as rectifiers (see later), although this term is normally used for diodes that can handle fairly high currents. The two terminals of a diode are called the *cathode* and the *anode* (often abbreviated to + and − respectively); the cathode is normally marked by a coloured band around the appropriate end of the component's body. This is shown in Figure 3.1, which also gives the circuit symbol for a diode. Figure 3.2 shows typical diodes.

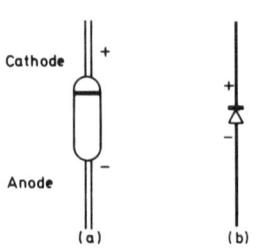

Figure 3.1 (a) Coloured band around the body of a diode indicates the cathode terminal; (b) circuit symbol for a diode

38

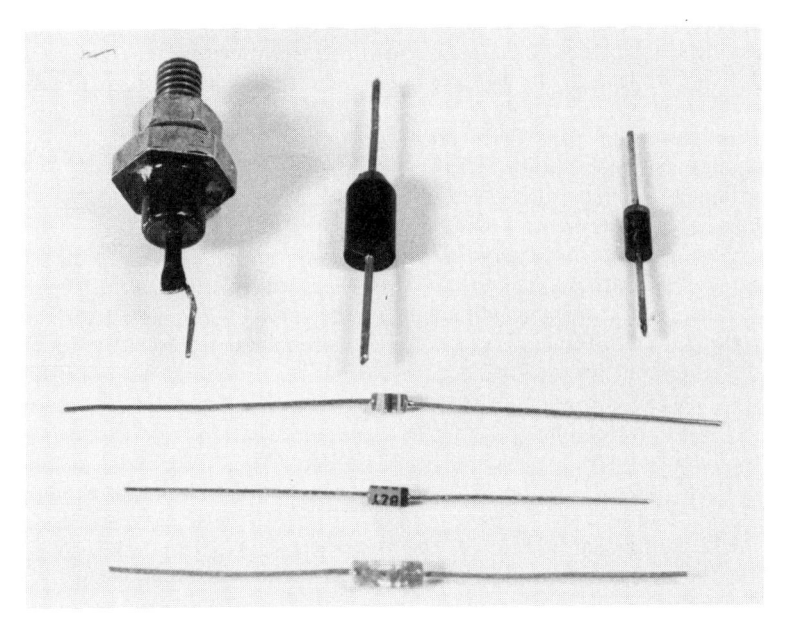

Figure 3.2 Top: three rectifiers (the one on the left can be bolted to a heatsink to conduct the heat generated within it away into the air). Bottom: a silicon diode, a zener diode and a germanium diode

If you construct the circuit of Figure 3.3(a) on the demonstrator unit, the diode will be found to conduct readily, as indicated by a large deflection of the meter. Reversing the polarity of the diode as shown in Figure 3.3(b) gives such a low current flow that it will not register on the meter at all. In terms of conventional current flow, the 'arrowhead' formed by the triangular part of the diode symbol shows the direction in which the diode passes a current.

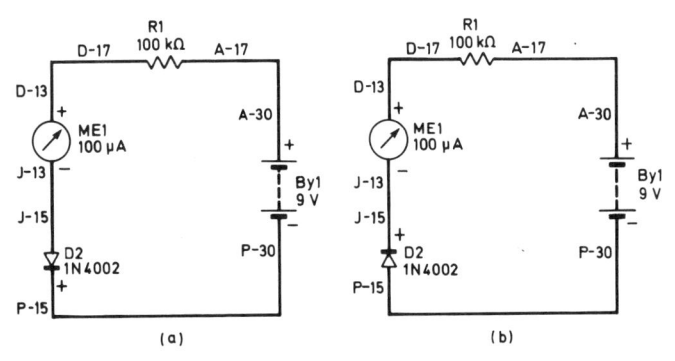

Figure 3.3 With the polarity shown in (a), the diode permits current flow; with polarity reversed, as in (b), no significant current flows

39

Ideally a diode would have an infinite reverse resistance and zero forward resistance, or perhaps more realistically a very high reverse resistance and a reasonably low and constant forward resistance. Practical diodes often have very high reverse resistances: often many thousands of megohms, provided that their peak inverse voltage (p.i.v.) rating is not exceeded. If this should happen, the resistance of the component falls dramatically and it can easily be destroyed.

In the forward direction the characteristics of semiconductor diodes are less satisfactory. *Silicon* diodes (such as the 1N4002 used in Figure 3.3) do not conduct significantly in the forward direction until the applied voltage reaches about 0.5 − 0.6 V. Increasing the voltage above this threshold level causes a large current to flow, with the effective resistance of the component dropping to a very low level. In the circuit of Figure 3.3(a) this effect shows up as an increase of about 0.5 V in the reading on ME1 (which effectively forms a 10 V f.s.d. meter in conjunction with R1) if the diode is replaced with a wire link. *Germanium* diodes have a lower forward voltage drop, and replacing D2 with an OA91 germanium device will clearly show this. The threshold voltage for germanium devices is about 0.1 V. However, these have the disadvantage of a relatively low reverse resistance, and connecting an OA91 into the circuit of Figure 3.3(b) may well result in ME1 indicating more than half f.s.d. Germanium devices are also relatively easily damaged by heat when they are being soldered into a circuit. Of course, the small voltage drop produced across a semiconductor diode is often of little or no importance, but it can give rise to problems in some applications.

Rectification

One of the main uses of diodes is with mains power supplies where the a.c. output from a step-down transformer must first be rectified to a

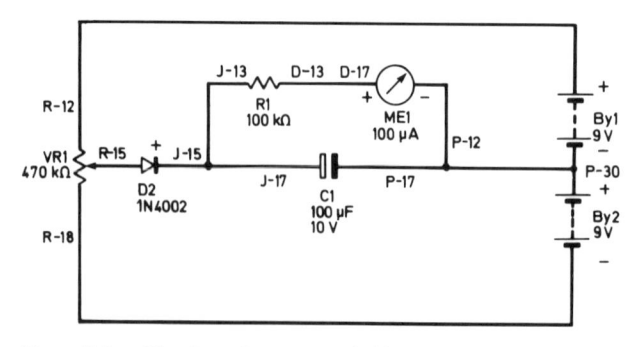

Figure 3.4   Circuit to demonstrate halfwave rectification

pulsating d.c. and then smoothed to a steady d.c. The simplest type of circuit for achieving this is a halfwave circuit such as that shown in Figure 3.4.

To avoid using the dangerous mains supply, this circuit enables you to generate an a.c. input signal using two batteries and potentiometer VR1 – the method described in Chapter 2 (Figure 2.16). If C1 is initially left out of the Figure 3.4 circuit, with VR1's slider adjusted to the lower half of its track, D2 will not conduct and no voltage will be fed to the voltmeter. If VR1's slider is brought up to the top of its track and then back down again, D2 conducts and gives a varying d.c. output during the period when VR1's slider is significantly more than half-way up its track. Thus the a.c. input waveform shown in Figure 3.5(a) emerges from the rectifier as a pulsating d.c., as shown in Figure 3.5(b).

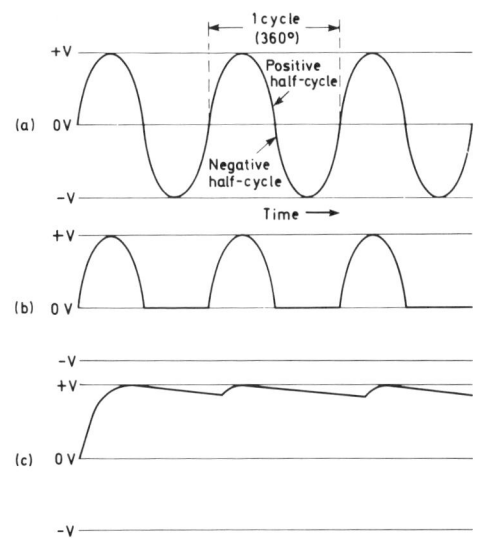

Figure 3.5 Feeding an a.c. waveform, (a), through a halfwave rectifier produces an output, (b), in which negative half-cycles have been eliminated; a smoothing capacitor can be used to fill the gaps between half-cycles and give a reasonably steady d.c. output, (c)

If capacitor C1 is then added to the circuit, this component stores a charge which tends to hold the output voltage at virtually its peak positive d.c. level during the intervals between positive peaks. Obviously C1 discharges to some extent during these intervals, and there is a small 'ripple content' on the output. However, provided that a suitably long time constant is produced by the smoothing capacitor and the load connected across the output, a reasonably steady d.c. will be obtained – see Figure 3.5(c).

Fullwave rectification

Smoothing the rectified d.c. output would be much easier if the negative half-cycles could be inverted so as to produce a constant stream of positive pulses, as shown in Figure 3.6. This gives the smoothing capacitor only half the previous discharge time between peaks, and therefore gives less ripple on the output for a given level of smoothing.

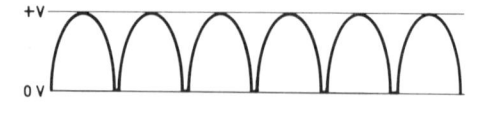

Figure 3.6 Fullwave rectification reduces the time between current peaks, and eases the problem of smoothing

One way of obtaining 'fullwave' rectification, as it is called, is to use the 'push-pull' arrangement of Figure 3.7. This requires a transformer with a centre-tapped secondary, or else two identical secondaries connected to give the same effect. When T1 provides D2 with a positive signal, D3 is fed with a negative signal. On the opposite half-cycles D3

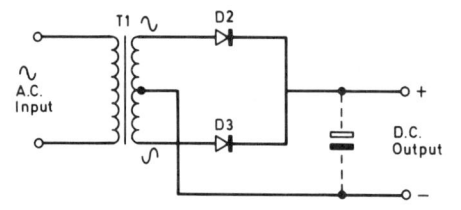

Figure 3.7 Fullwave push-pull rectifier

receives the positive signal and D2 is fed with the negative one. Thus D2 and D3 alternately provide positive output pulses, and the required full-wave rectification is obtained. The signals fed to D2 and D3 are said to be *in antiphase*, or 180 degrees out of phase. This refers to the fact that in one complete cycle, when drawn graphically as in Figure 3.5(a), the line goes through 360 degrees, 180 degrees in each half-cycle. If two signals are half a cycle out of step they can therefore be considered to be 180 degrees out of phase. Signals that are precisely in step with one another are said to be *in phase*.

An alternative form of fullwave rectifier is the bridge circuit. This requires four diodes, but needs only a single untapped secondary winding on the transformer. The arrangement is shown in Figure 3.8.

When T1 provides a positive voltage at the top end of the secondary, current flows through D3, the load connected across the output, and then through D4 to the negative end of the secondary. When the

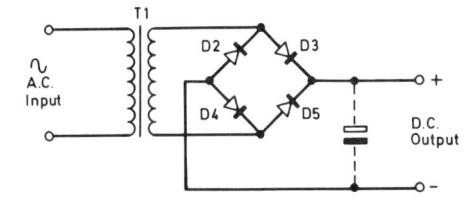

Figure 3.8 Fullwave bridge rectifier

secondary polarity is reversed, current flows through D5, the load, and then through D2 to the negative side of the secondary. Effectively the secondary is connected to the output first one way round, then the other, so that it always provides a positive output.

Note that if the diodes in these rectifier circuits are all connected the opposite way round, the circuits still work, but the polarity of the d.c. output will be reversed.

Zener diodes

A zener diode is used to provide a stable voltage from a potential that tends to fluctuate somewhat. In the case of a mains power supply small changes tend to occur in the mains voltage. With a battery the voltage tends to drop as it gradually becomes discharged, the actual output from a 9 V battery dropping from about 9.5 to 8 V during its working life. The internal resistance of the supply also causes voltage changes, with the output potential dropping with increased output current. 'Internal resistance' is merely the resistance that must inevitably occur within the internal structure of a battery or mains power supply. (In the case of the latter, for instance, it is largely composed of the resistance through the secondary winding of the transformer and the rectifiers.) This resistance produces a voltage drop, in much the same way as if it were an actual resistor connected in series with the output of the supply. Thus if a battery has an internal resistance of 20 ohms, drawing a 100 mA current causes the output voltage to drop by 2 V ($20 \times 0.1 = 2$).

A zener diode can be regarded as being much the same as an ordinary silicon type, but its reverse breakdown voltage is known quite accurately. In use it is connected across the unstabilised supply via a current-limiting resistor, in the manner indicated in Figure 3.9. Note that the circuit symbol for a zener is slightly different from that for an ordinary diode. Figure 3.10 shows the circuit set up on the demonstrator unit.

Although the zener is reverse-connected, it conducts because the supply voltage is greater than its reverse breakdown voltage, causing it to 'avalanche'. R2 limits the current through the device to a safe level.

43

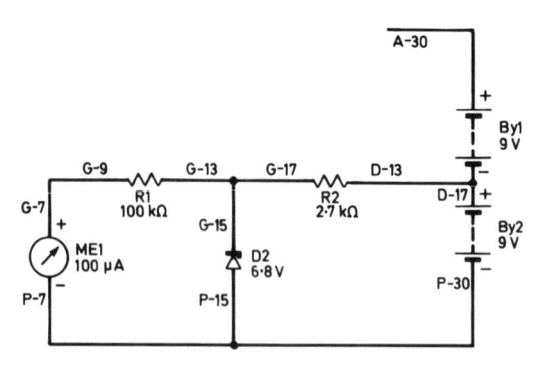

Figure 3.9 Circuit to demonstrate the use of a zener diode as a voltage stabiliser

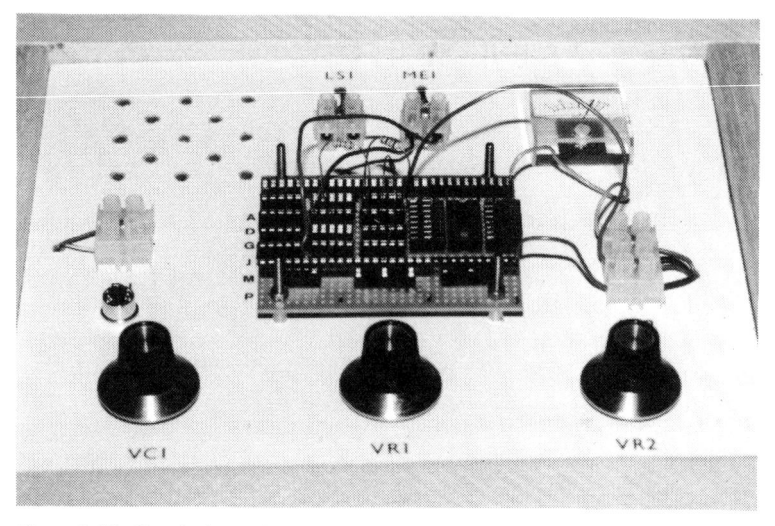

Figure 3.10 Circuit shown in Figure 3.9 set up on the demonstrator unit

D2 has a voltage rating of 6.8 V, and within a few per cent this is the reading that should be obtained on ME1. If the right-hand terminal of R2 is disconnected from point D-13 and reconnected at A-13, the zener circuit is then fed from double the previous supply voltage. The avalanche effect of D2 still limits the voltage across D2 to about 6.8 V, and the reading on ME1 will probably not noticeably increase. If an output current is extracted from the circuit by connecting a resistor across D2, the meter reading will still say quite stable (provided the voltage-divider effect across R2 and the load resistor R1 does not drop the voltage fed to D2 to less than 6.8 V). In a practical circuit the value of R2 must be made low enough to give the required output current without dropping the voltage to the zener below its rated voltage, and the zener must be capable of handling the power developed across it with no load connected at the output.

For low zener voltages it is common to use a forward-connected silicon diode. The forward threshold voltage gives a zener voltage of about 0.65 V, and this can be demonstrated by reversing the polarity of D2 in Figure 3.9. Voltages of about 1.3 V and just under 2 V can be obtained by using two and three diodes respectively, series-connected.

Light-emitting diodes

Some form of radiation is produced by all forward-biased semiconductor diodes, and light-emitting diodes (LEDs) are designed to radiate as much energy in the visible light spectrum as possible (infra-red types are also available). The main advantage of an LED over a small filament bulb is its far lower power requirement. Many LEDs will produce a reasonably bright light from an input of only about 2 mA at less than 2 V, whereas a small light bulb would probably need something like 40 – 60 mA at 6 V (nearly one hundred times as much). An LED achieves high efficiency, as it wastes very little power in the form of unwanted heat.

An LED cannot be used just like an ordinary light bulb. For a start it must be *forward*-biased before it will produce light, and the supply voltage is very critical. At fractionally below the conduction threshold, no significant current will flow and no light will be produced. Slightly above this voltage an excessive current results and the diode is destroyed.

In practice the LED is simply fed from a supply of three volts or more, and a series resistor is used to prevent an excessive current flow. In effect the LED acts as a zener that stabilises its supply voltage at the correct level. If we wish to run an LED at 5 mA from a 9 V supply, for example, we first deduct 2 V from the supply potential (the approximate voltage developed across the LED) to obtain the voltage developed

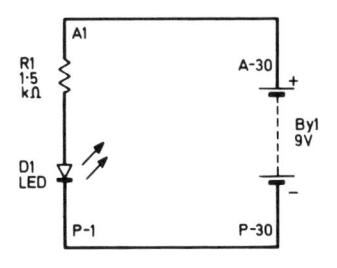

Figure 3.11 Circuit to demonstrate the use of an LED

across the resistor. We thus have figures for current and voltage, and the appropriate resistance can then be calculated using Ohm's law ($R = E/I = 7/0.005 = 1400$ ohms). This resistance value is not available, so the nearest available value of 1.5 kΩ must be used; hence we have the circuit shown in Figure 3.11 (note the modified circuit symbol for the LED). Incidentally, these available values are known as 'preferred values', and although the sequence of values may at first sight seem to be a rather strange progression, closer examination will show that it does actually give a reasonably uniform progression in percentage terms.

Transistors

Unlike the components described so far, which have all been passive types, the transistor is an 'active' device, i.e. it can provide amplification. It is also different in that it has three and not two leadout wires. These are called the *base, collector,* and *emitter* terminals for a bipolar transistor (the type in most common use; there are others such as the field-effect type and the unijunction type, but these will not be considered just yet). There are actually two types of bipolar transistor:

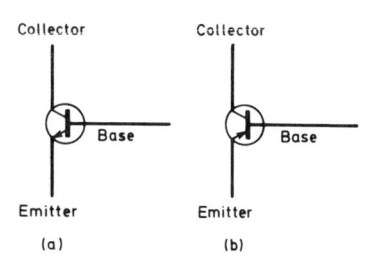

Figure 3.12 Circuit symbols for bipolar transistors: (a) npn, (b) pnp

Figure 3.13 Selection of transistors. Top row: two power transistors (which can be bolted to a heatsink to conduct the heat generated within them away into the air) and a germanium transistor. Bottom row: four silicon transistors

npn and pnp. An npn device is so called because it is constructed from a piece of p-type semiconductor material sandwiched between two pieces of n-type material. Similarly, a pnp device is constructed from a piece of n-type semiconductor material sandwiched between two layers of p-type material. The circuit symbols for npn and pnp bipolar transistors are shown in Figure 3.12; you will see that the only difference between the two is the direction of the arrow in the emitter part of the symbol. Figure 3.13 shows some typical transistors.

The circuits that follow are mainly based on the BC109 npn silicon transistor. They could employ the BC179 device, which is the pnp equivalent of the BC109, but the supply polarity would need to be reversed. This is the only practical difference between the two types. It should perhaps be pointed out that transistors, like diodes, can be based on silicon or germanium, but germanium devices are generally inferior to silicon types, and are rarely used these days. Thus in what follows we shall be concerned with the features of silicon devices (germanium devices are slightly different in some points of detail).

Current gain

The circuit of Figure 3.14 is for a sort of simple transistor tester, and it also shows that a transistor can provide current amplification. Initially

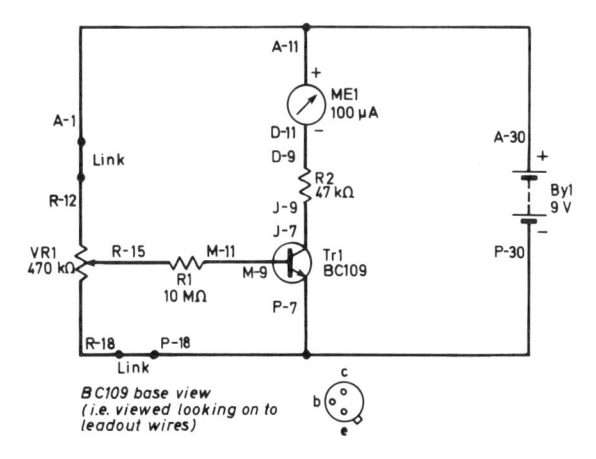

Figure 3.14 Circuit to demonstrate the current-amplifying property of a transistor

the slider of VR1 should be at the bottom of its track. In theory there should be no current flow through ME1, R2, and the collector–emitter junction of Tr1. A small leakage current does actually flow in practice, but for a silicon device this should be too small to register on the meter.

47

Taking VR1's slider up its track applies a forward bias to the base—emitter junction of Tr1, and a significant current begins to flow in this part of the circuit when the bias voltage rises above about 0.6 V. This junction acts as a silicon diode in fact, and in this case it is being forward biased. The forward bias causes a current to flow through the collector—emitter terminals of Tr1, causing a deflection on ME1. Furthermore, this collector—emitter current varies roughly in proportion to the level of current fed into the base—emitter junction. ME1 should reach full scale deflection with VR1 adjusted for a slider potential of only about half the supply potential.

This may not seem of great importance until you calculate from Ohm's law the base current of Tr1. It then becomes apparent that a maximum base current of less than 1 µA can flow via R1 into Tr1. In other words, a variation of less than 1 µA into Tr1 base gives a change in collector current of 100 µA, so the transistor is providing a current gain of over one hundred times. The static current gain of a transistor (normally abbreviated to h_{FE}) is given by:

$$h_{FE} = \frac{\text{collector current}}{\text{base current}} = \frac{I_c}{I_b}$$

The current gain of the BC109 device is actually 240 to 900 times, at a collector current of 2 mA and a collector voltage of 5 V. It is necessary to specify the collector current and voltage since these both affect the gain of a transistor. The gain of a normal transistor tends to drop considerably at low levels of collector current, and it usually falls to a lesser extent at very high collector currents. Provided the collector voltage is not less than about one volt, it has little affect on gain, although gain does rise very slightly with increasing voltage. Reducing the collector voltage below about one volt causes the gain level to diminish rapidly.

Ratings

Transistors have a number of important parameters, of which current gain is just one example. Some of the others are listed below.

P_{tot} maximum power dissipation in free air
I_c maximum permissible collector current
V_{ce} maximum permissible collector-to-emitter voltage
V_{cb} maximum permissible collector-to-base voltage
V_{eb} minimum reverse breakdown voltage of the base—emitter junction.

Biasing

In order to use a transistor in a practical amplifier circuit it is necessary to provide a suitable input bias current and an output load resistor. The simplest type of practical amplifier uses the arrangement shown in the demonstration circuit of Figure 3.15. R2 is the load resistor and R3 is the bias resistor. The other components are needed to demonstrate the properties of this arrangement, as we shall see shortly.

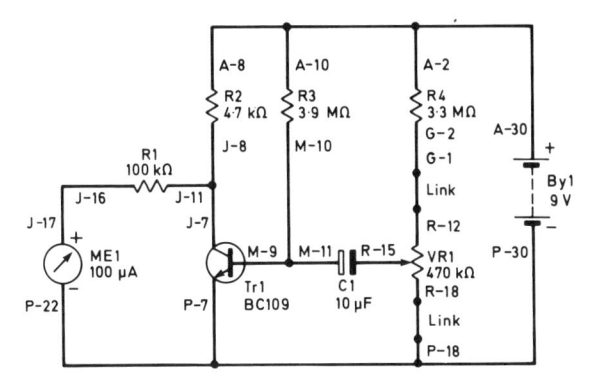

Figure 3.15 Circuit to demonstrate simple biasing

The value given to R2 depends upon the chosen static collector current for Tr1, and the supply voltage. In this case a 1 mA collector current has been chosen, as this is low enough to give low battery drain and dissipation in the components, but high enough to give good gain from Tr1 (in practice the level of collector current is rarely critical). The value of R2 is chosen to give the required static collector current when it is subjected to half the supply voltage, and from Ohm's law this gives:

$$R = \frac{E}{I} = \frac{4.5}{0.001} = 4500 \text{ ohms or } 4.5 \text{ k}\Omega$$

Therefore R2 is given the nearest preferred value of 4.7 kΩ.

R3 should have a value that gives the required static collector current through Tr1. The base current required to do this is equal to the h_{FE} of Tr1 divided into the collector current: 0.001/450 = 0.0000022 A = 2.2 μA. The voltage across R3 is approximately equal to 9 V (the supply voltage) minus 0.6 V (the voltage dropped across Tr1 base–emitter), i.e. 8.4 V. From Ohm's law we can then calculate the required value for R3:

$$R = \frac{E}{I} = \frac{8.4}{2.2} = 3.8 \text{ M}\Omega$$

(The answer is in megohms because the current was entered into the equation in µA, not amps, in order to simplify the calculation.) R3 is given a value of 3.9 MΩ in practice, as this is the nearest preferred value to 3.8 MΩ.

If you try out the circuit, it is possible that the voltage reading obtained from ME1 will confirm that there is approximately the designed potential of 4.5 V at Tr1 collector, and that the collector current is actually about 1 mA therefore. However, it is quite likely that it will be considerably off-target. This is partly due to the tolerances of R2 and R3, and the fact that their nominal values are slightly different from the calculated values anyway. To a larger extent it is due to the tolerance in the gain of Tr1, which in practice could be anything from about 200 to nearly 900 in this circuit. With a particularly high-gain device the collector voltage of Tr1 will be at virtually zero, whereas a low-gain device would give a collector potential approaching 9 V.

Clipping

For the reasons just given, a simple circuit such as this has rather unpredictable levels of performance, particularly with regard to the maximum peak-to-peak output voltage swing. This can be explained by coupling a small a.c. signal into Tr1's base by adjusting the control knob of VR1 to and fro over a few degrees of rotation. Positive input signals cause Tr1 to conduct more heavily and its collector voltage to fall; negative input signals have the opposite effect. This is indicated by fluctuations in the reading on ME1. If VR1 is adjusted by a suitably large amount, the difference between the output voltage on negative and positive signal peaks (i.e. the peak-to-peak output voltage) will be

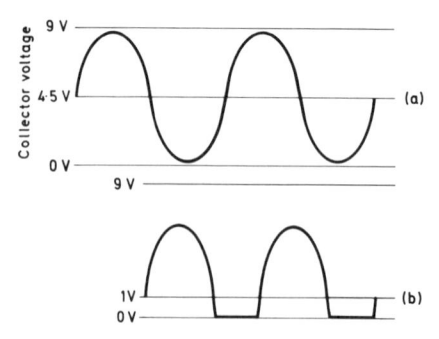

Figure 3.16 (a) With quiescent collector potential of 4.5 V, an unclipped output signal having a peak-to-peak value virtually equal to the supply voltage can be obtained. (b) With the quiescent collector voltage well away from the 4.5 V level (1 V in this example), trying to obtain a 9 V peak-to-peak output signal results in the signal being clipped

virtually equal to the supply potential, as shown in Figure 3.16(a). This assumes that the required 4.5 V quiescent collector voltage is obtained. If the quiescent voltage is (say) 1 V, then on negative output signals Tr1 would be unable to go sufficiently negative, and would severely distort the signal in the manner shown in Figure 3.16(b). This is termed *clipping*.

Of course, even with the quiescent output voltage at 4.5 V, a high input signal could still drive the circuit to the point where the output signal clips symmetrically, but it does at least give the highest possible overload margin. Similarly, even with the biasing well off-centre, a small output signal level may not cause clipping. However, higher distortion and a generally inferior level of performance would probably still be obtained.

Negative feedback

A more stable form of biasing is shown in Figure 3.17. The top end of R3 is now taken to Tr1 collector instead of to the positive rail, and its value has been made slightly less than half its previous level. This is simply because it is now subjected to only a little under half the voltage it received in the previous circuit, but must provide the same current level.

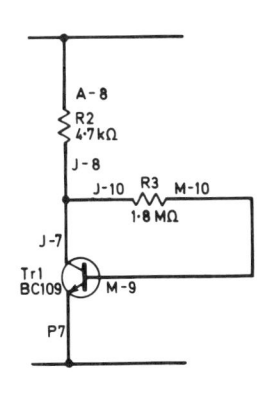

Figure 3.17 Amendment to the Figure 3.15 circuit: negative feedback gives a more stable form of biasing

If Tr1 has a particularly high gain, its collector voltage tends to go below 4.5 V; this reduces the voltage across R3, reduces the bias current, and increases the collector voltage of Tr1. If Tr1 is a low-gain device, its collector voltage tends to go above 4.5 V, giving increased bias current, and reducing the collector voltage. Thus the circuit has a stabilising *negative feedback* action, which gives more predictable results.

51

Potential-divider bias

The above method gives reasonably reliable results, but in some circumstances an even more reliable method is needed, and this can be achieved using the type of circuit shown in Figure 3.18.

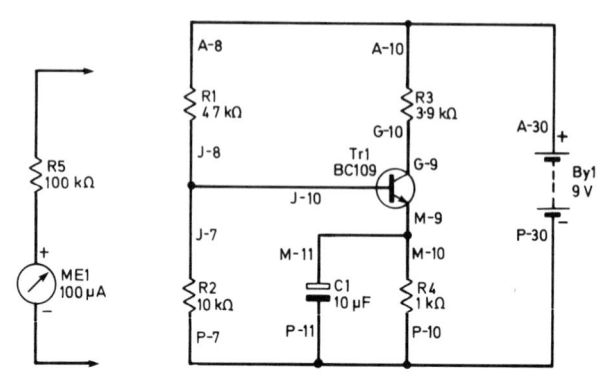

Figure 3.18 A highly stable biasing arrangement

R1 and R2 form a voltage-divider circuit that provides a nominal 1.6 V at Tr1 base. The value of R2 must be made quite low in relation to the input resistance of Tr1 so that Tr1 has no significant shunting effect on R2 and the voltage produced by this network. There is about 0.6 V dropped across the base—emitter junction of Tr1, giving 1 V across R4 and a consequent current flow of 1 mA. The emitter and collector currents of a transistor are virtually identical (the emitter current is fractionally the higher of the two, since it consists of the collector current plus the comparatively insignificant base current). Thus there is a current of approximately 1 mA flowing through R3, and about 3.9 V is developed across this component. This leaves about 4.1 V across the collector—emitter terminals of Tr1 (as calculated using Kirchhoff's law). This permits an unclipped peak-to-peak output voltage of about 7.8 V (± 3.9 V), and reliably sets the operating current of Tr1. If you set up the circuit on the circuit demonstrator unit, and use the voltmeter consisting of ME1 and R5, you should be able to obtain readings within a few per cent of the specified voltages.

Capacitor C1 is needed to remove the large amount of negative feedback that would otherwise reduce the voltage gain of the amplifier to only about 3.9 times. In the Figure 3.17 circuit there was a fairly obvious feedback loop through bias resistor R3, but there is no apparent feedback path in this circuit, and this often puzzles the beginner. In fact the negative feedback is applied from Tr1's emitter to its base through the structure of the transistor. If a positive input signal is applied at Tr1 base, for example, the base voltage will obviously increase, but so will the emitter voltage because it is only loosely coupled to the negative rail via R4. This causes an increase in the voltage across R4, and therefore produces an increase in the current through R3, Tr1's collector—emitter terminals, and R4. The current increases through R3 and R4 are virtually identical, but the voltage increase across R3 is 3.9 times higher than that across R4, owing to the higher value of the former. Thus the voltage gain is only some 3.9 times.

Provided the time constant of C1 and R4 is suitably large, C1 holds the emitter voltage of Tr1 at its quiescent level even in the presence of an input signal, and the feedback is thus eliminated.

Modes of operation

So far we have only considered the operation of the transistor in one amplifying mode, the *common-emitter* mode. It is so called because the emitter terminal is common to both the input and output signal, as can be seen from Figure 3.19(a). A transistor can also operate as an amplifier with the collector or the base as the common terminal, and these basic configurations are shown in Figure 3.19(b) and (c).

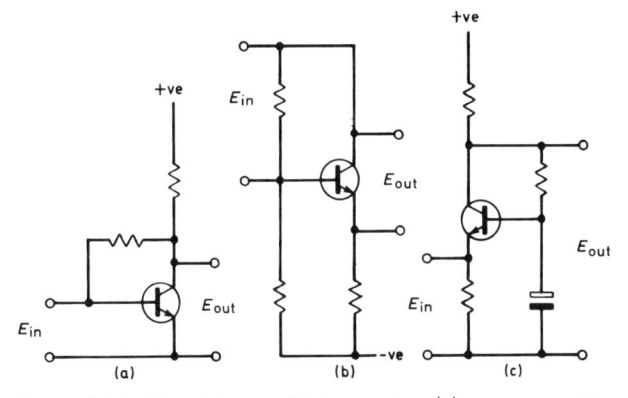

Figure 3.19 Transistor amplifying modes: (a) common-emitter, (b) common-collector (emitter follower), (c) common-base

Each mode of operation gives different levels of performance with regard to voltage gain, input impedance, and output impedance. The first of these terms is really self-explanatory, and is simply the output voltage change divided by the applied input voltage change. When an input voltage is applied to an amplifier, an input current flows, rather in the same way as when a voltage is applied to a resistor. The input impedance of an amplifier could therefore be regarded as the resistance it provides to an a.c. input. As stated in Chapter 1, impedance is equal to peak voltage divided by peak current, and is used when a circuit has

53

some combination of resistance, capacitive reactance and inductive reactance. The input impedance of a transistor amplifier is mainly resistive, with a small parallel capacitive element, and so it reduces somewhat at high frequencies.

If the output from an amplifier is coupled to a load resistor, the output voltage from the amplifier drops, owing to a potential divider action between the load resistor and resistances in the output circuitry of the amplifier. The effect is rather as if there were a resistor in series with the output from the amplifier, with a voltage drop occurring across this resistor when an output current is drawn. The output impedance of the amplifier is equal to the value of this effective series resistor. Owing to small capacitive and inductive elements in the circuitry, the output impedance does actually vary to some extent with input frequency.

It should perhaps be briefly explained that impedance is defined as *peak* voltage divided by *peak* current (rather than simply voltage divided by current) because when an a.c. signal is applied to a circuit having inductance and/or capacitance, peak current and voltage levels do not necessarily coincide. However, the theory of this goes beyond the scope of this book.

Common-emitter characteristics

The common-emitter amplifier is the type most often encountered, as it has what is generally the most useful set of characteristics. The input impedance is usually in the region of a few kilohms, as is the output impedance, and voltage gains of more than 100 times can readily be obtained. For most practical purposes the output impedance can be regarded as equal to the resistance of the collector load resistor.

At audio frequencies the input impedance is approximately equal to:

$$\left(\frac{25}{\text{collector current in mA}}\right) h_{FE}$$

Thus it is about 11.25 kΩ for the circuit of Figure 3.15, for example. If there is an emitter resistor that is not bypassed by a capacitor, this should be added to the answer obtained from the first part of the formula before multiplying by the h_{FE} value. The input impedance can be considerably boosted in this way. With the potential-divider method of biasing, the input signal causes a current to flow in the base bias resistors, and they are effectively connected in parallel with the input impedance of the transistor. Their shunting effect must therefore be taken into account, and significantly reduces the input impedance of the amplifier in most cases.

The voltage gain is roughly equal to:

$$\frac{\text{collector load resistance}}{25 \div \text{collector current in mA}}$$

This comes to about 188 times for the circuit of Figure 3.18. If an unbypassed emitter resistor is used, this must be added to the answer from the lower line of the formula before it is divided into the collector load resistor value.

Common-collector characteristics

This type of circuit is more popularly known as the 'emitter follower' configuration. It provides a voltage gain of slightly less than unity, as the emitter voltage is free to rise and fall in sympathy with changes in the base voltage. However, it is still a very useful form of circuit, as it realises the full current gain of the transistor, and can therefore be used as a buffer stage to either reduce the output impedance or increase the input impedance of an amplifier.

The output impedance of an emitter follower is approximately equal to the output impedance of the stage driving its input divided by the current gain of the transistor. The emitter load resistor must have a value that is low enough to give a quiescent current at least equal to the required peak output current.

The input impedance is roughly equal to $h_{FE}(r_e + R_e)$, where r_e is equal to 25 divided by the collector current in mA, and R_e is the value of the emitter load resistor. Note that the input impedance of the stage driven by the circuit is effectively in parallel with R_e, and reduces the input impedance. Also, the two base bias potential-divider resistors are effectively in parallel with the input impedance, and will have a considerable shunting effect on it.

Bootstrapping

The shunting effect of the bias resistors can be greatly reduced using the bootstrapping technique — which is employed in a number of applications. In Figure 3.20 Tr1 is biased from the potential divider (R1—R2) via series resistor R3, and some of the output signal is coupled to the junction of these three resistors by C1. If an input signal takes Tr1 base (say) 1 V more positive, then the output also goes 1 V positive, as does the junction of R1—R2—R3 because of the coupling through R3. Thus, although the right-hand side of R3 has increased in voltage, the other terminal has increased by the same amount, and so the input signal does not cause a current to flow in R3. It has an apparent infinite

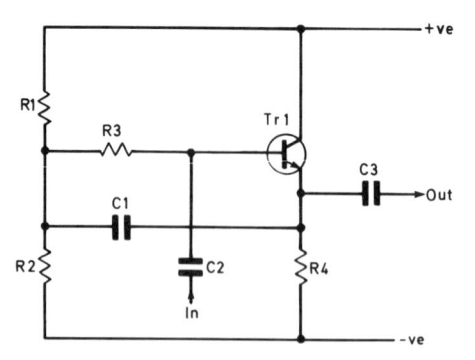

Figure 3.20 The bootstrapping technique can be used to reduce the shunting effect of the bias resistors on the input impedance of an emitter-follower stage

impedance to the input signal, and therefore does not shunt the input impedance.

In practice there is a slight shunting effect, because Tr1 has a voltage gain of very slightly less than unity (about 0.98 to 0.99), and a very small amount of input current does flow through R3. The shunting effect is greatly reduced, though.

Common-base mode

In a common base amplifier the base is forward biased in much the same way as for a common-emitter stage. However, the base—emitter voltage of the transistor is varied by holding the base at a fixed potential and applying the input signal to the emitter, rather than the other way round as in a common-emitter stage. Either the base potential can be fixed using a potential divider network having suitably low values, or a 'bypass' capacitor can be used to smooth out any variations in base potential.

Like a common-emitter stage, a common-base amplifier has an output impedance roughly equal to the collector load resistor value, but this is where the similarity ends. The voltage gain is slightly higher than for a common-emitter stage, having a comparable collector load value and collector current, but there is approximately unity current gain through the stage. This gives a very low input impedance, which is roughly equal to the collector load value divided by the level of voltage gain. In practice this normally works out at only about 100 to 200 ohms, and together with the low power gain of this configuration explains why it is so little used.

It is worth noting that a positive input to a grounded base stage causes a reduction in the base—emitter voltage and the collector current.

This gives a rise in the output voltage, so that the input and output signals are in phase (as they are with the emitter-follower mode also). As we saw in the experiment of Figure 3.15, the signal is inverted by a common-emitter stage.

Microphone preamplifier

Unfortunately it is not possible to verify the facts and figures given about transistor operating modes without going into a lot of fairly complex theory, or making practical tests using sophisticated test equipment, neither of which would be appropriate here. However, making up the microphone preamplifier circuit of Figure 3.21 will permit a few simple tests to prove that the basic characteristics of the various amplifying modes are as stated earlier.

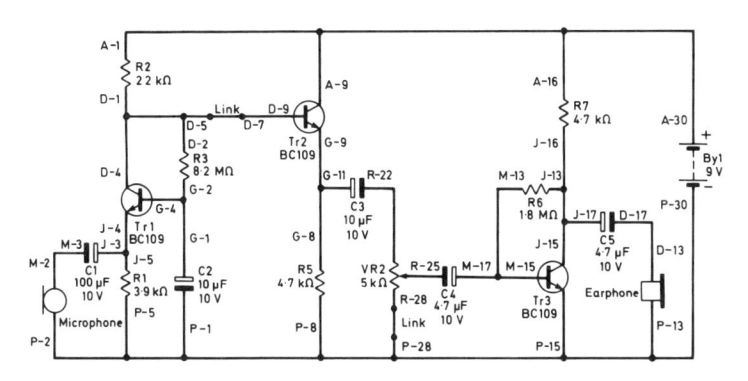

Figure 3.21 A microphone preamplifier incorporating transistors in all three amplifying modes

A low-impedance dynamic microphone has only a very low output voltage, but can provide a relatively high current. The microphone is therefore coupled to the input of a common-base stage, which gives considerable voltage amplification. Capacitor C1 prevents a current flowing from Tr1 emitter into the microphone, but allows the a.c. signals from the microphone to effectively pass to Tr1. The value of a coupling capacitor such as this should be chosen so that its reactance is somewhat less than the input impedance of the stage it feeds into at the lowest frequency to be coupled (about 30 Hz in the case of an audio circuit). The voltage drop across the capacitor, due to a sort of potential-divider action in conjunction with the input impedance of the following stage, then causes little attenuation at any frequency the circuit is designed to handle.

An emitter-follower buffer stage is used at the output of Tr1. This enables the relatively high output impedance of Tr1 to drive the much lower input impedance of the following circuitry with little drop in signal level. Tr2 is biased by the collector voltage of Tr1, and this is called 'direct coupling'. The output from Tr2 is coupled to a potentiometer, which operates as a volume control, and from here the signal is coupled to a common-emitter stage based on Tr3, which gives ample drive to provide good volume from a crystal earphone.

The unit can be fed from a proper low-impedance dynamic (cassette-recorder type) microphone, but a simpler alternative is simply to couple the loudspeaker on the circuit demonstrator to the input of the unit. This will then work in reverse as a sort of crude moving-coil microphone (a technique often used in intercoms).

Tests

The voltage gains of the three stages can be roughly gauged by disconnecting the earphone lead from point D-13, and connecting it in sequence to M-3, D-1, G-9, and J-15. These moves should give, respectively, too low a signal level to be detected, a much amplified but still low-level signal, a virtually identical signal level, and finally a very high signal level (controllable by means of VR1).

The low input impedance of Tr1 can be confirmed by adding a 4.7 kΩ resistor in series with the microphone. The potential-divider action across this resistor and the low input impedance will give a severe loss of signal. In contrast, replacing the link wire between points D-5 and D-7 with this resistor will give no noticeable loss of signal, because of the very high input impedance of Tr2. Connecting the resistor temporarily in series with C4 will give a readily detectable but not very substantial loss of signal, because of the medium input impedance of Tr3.

The low output impedance of Tr2 can be demonstrated by adding a resistor of about 470 ohms across the volume control: this should give only a small loss of signal. Connecting it across the earphone should produce a far larger loss of gain, because of the higher output impedance of Tr3, and there will be an even larger loss of volume if it is connected from Tr1 collector to the negative supply (via a 2.2 μF d.c. blocking capacitor), since Tr1 has an even higher output impedance.

Log and lin

A logarithmic (log) potentiometer is used for the volume control; indeed this is the main application for log types. A log potentiometer differs from a linear (lin) type in that moving the slider up the track does not produce a steady and regular increase in slider voltage with the former, but does with the latter. This 'logarithmic law' can be clearly shown using the test circuit of Figure 3.22. For about the first 180 degrees of rotation, the control knob of VR2 will have little effect on the output

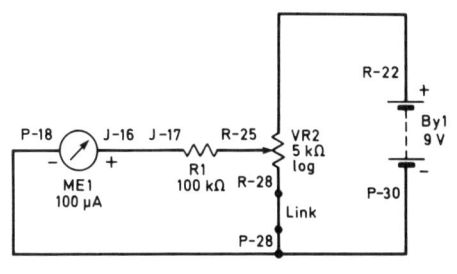

Figure 3.22 A log potentiometer does not provide a steady and regular change with the slider voltage, but gives a greater change with the slider operating near the top end of the track; this gives more acceptable results than a lin type in volume-control applications

voltage, but in the last 90 degrees there will be a substantial increase in the output potential. The characteristic of human hearing is such that a logarithmic potentiometer seems to give a smooth and even control of volume, whereas a linear type does not.

Class B amplification

The method of biasing described earlier is known as *class A* biasing, and is the form of biasing that is almost invariably used in low-power stages. It is less commonly used in output stages, because it requires a quiescent current flow equal to the peak output current required; in higher-power stages this results in a continuous and large current consumption, and a high dissipation in the transistor and its load resistor. It would be preferable to have a circuit that had a low quiescent current, with the current drain increasing in proportion to the output power.

This can be achieved using a *class B* amplifier, such as the simple audio power-amplifier circuit shown in Figure 3.23. Tr1 and Tr2 are used as a common-emitter driver stage, and they are connected to form what is called a 'Darlington pair'. This merely consists of coupling the output current of the first transistor into the base of the second transistor, which effectively produces a transistor having an h_{FE} value equal to the product of the h_{FE} values of the individual transistors. This does not actually increase the sensitivity of the amplifier, since the voltage gain of the circuit is controlled by bias resistors R1 and R2 (this is explained in Chapter 4, dealing with operational amplifiers), but it produces increased negative feedback and thus improved quality, as

59

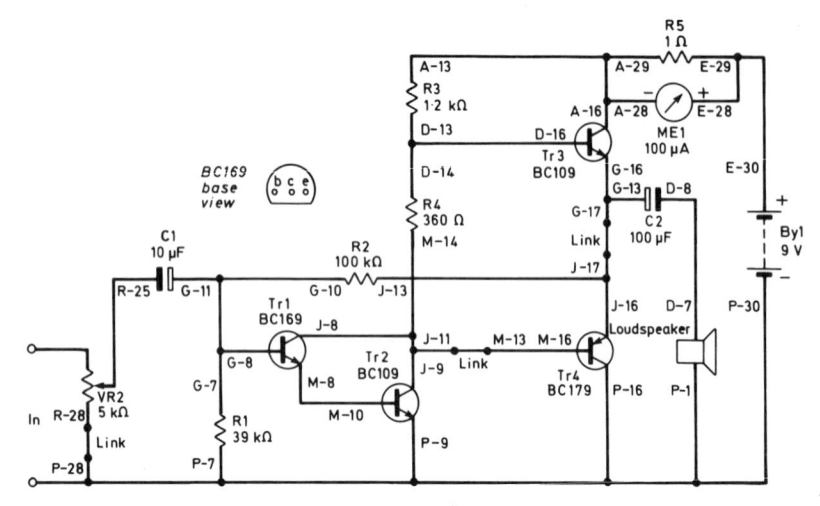

Figure 3.23 Simple class B amplifier circuit (note: the BC169 is electrically the same as the BC109 but has a different leadout arrangement, which is more convenient for the Tr1 position if the circuit is constructed on the demonstrator unit)

explained later. The threshold voltage of a Darlington pair is double that of a single transistor (i.e. about 1.2 V), and so R1 and R2 have values that produce 1.2 V at Tr1 base with about half the supply voltage at the output of the amplifier (Tr3 and Tr4 emitters).

Quite a high current is needed to drive even a small high-impedance loudspeaker, and an emitter-follower output stage is used to provide a suitably low output impedance. In fact two emitter-follower transistors are used, npn device Tr3, and pnp device Tr4, forming a complementary arrangement. The value of R4 is chosen to give a little under 1.2 V across the bases of Tr3 and Tr4 so that they pass a small quiescent current. R3 is the main collector load for Tr1–Tr2.

With an input signal applied to the circuit, on positive-going output signals Tr4 plays no active role, but Tr3 conducts heavily, causing a charge current to be supplied to C2 through the speaker. The larger the voltage swing, the higher this charge current, and the greater the voltage developed across the speaker. On negative-going output signals Tr3 plays no active role, but Tr4 becomes conductive and discharges C2 via the speaker. The more negative the output swings, the higher the discharge current, and the greater the voltage developed across the speaker. This effectively couples the voltage swing at Tr3 and Tr4 emitters to the loudspeaker, and gives high efficiency, since the current drain varies in proportion to the output power.

This last point will be clearly shown by the reading on ME1, which varies from little more than zero under quiescent conditions to probably well beyond half f.s.d. at high volume levels (a suitable input signal can be extracted from the earphone socket of a portable radio or cassette recorder). ME1 is actually far too sensitive to measure the quite high currents drawn by the amplifier, so R5 has been added in parallel to reduce its sensitivity. R5 has a value that is only about one-thousandth of the resistance through ME1; therefore the voltage developed across these causes approximately 1 mA (1000 μA) to flow through R5 for every μA that flows through ME1. This gives the current-meter circuit a sensitivity of roughly 100 mA f.s.d. In fact R5 is what is known as a 'shunt resistor'.

Distortion

The standing bias provided by R4 is very important, because without it a voltage swing of ±0.6 V would be needed at Tr1–Tr2 collectors before any variation in output voltage would be produced at all! This would result in part of the output waveform being absent, as shown in Figure 3.24, and would give a severely distorted output. This is known

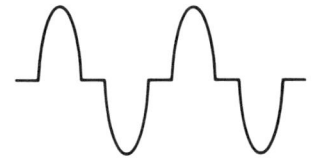

Figure 3.24 Crossover distortion results in part of the output waveform being absent

as *crossover distortion*, and becomes more severe at low output levels where the ±0.6 V becomes a more substantial fraction of the signal level. Of course, if R4 is made too low in value, it will not provide the full bias voltage required, and a small but significant degree of crossover distortion will still be present (as can be demonstrated by reducing R4 to, say, 220 or 270 ohms).

It is just as important that R4 is not made too high in value, as this would produce quite a high quiescent current. Furthermore, this would cause the output transistors to heat up, and increased temperature results in a reduction of the threshold voltage of a transistor (or diode). The effect would be the same as increasing the bias voltage, and would cause an increase in the quiescent current. There would then be an increased heating effect followed by a further rise in current. This could easily continue until the dissipation in the transistors was so high that they were destroyed by overheating. The effect is known as *thermal runaway*.

Crossover distortion is not the only form of distortion. All amplifiers, including class A types, produce distortion because of the variations in gain that occur with changes in the collector current and voltage of a transistor. Some parts of the input waveform receive more amplification than others, producing a distorted output waveform. Negative feedback tends to reduce distortion. It does this by reducing the voltage gain of the circuit to a level largely determined by two resistors, and influenced to a lesser degree by the innate voltage gain of the amplifier. Variations in this innate voltage gain therefore have comparatively little effect on the actual voltage gain, and reduced distortion results.

Distortion tends to be a severe problem with power amplifiers, owing to the large swings in output voltage and current. It is therefore usually necessary to use a large amount of feedback in order to obtain good results (i.e. the voltage gain set by the resistors must be much lower than the intrinsic voltage gain of the amplifier). This is why a high-gain Darlington pair driver stage is used in the circuit of Figure 3.23. The voltage gain of the circuit is actually only about four times.

Thyristors

The thyristor (also called a silicon controlled rectifier or s.c.r.) is a three-terminal device with the circuit symbol of Figure 3.25 (a). A typical

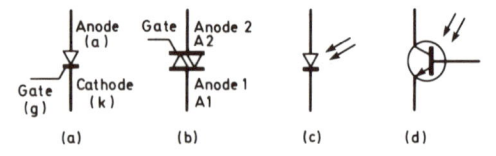

Figure 3.25 Circuit symbols for (a) thyristor (s.c.r.), (b) triac, (c) photodiode, (d) phototransistor (npn)

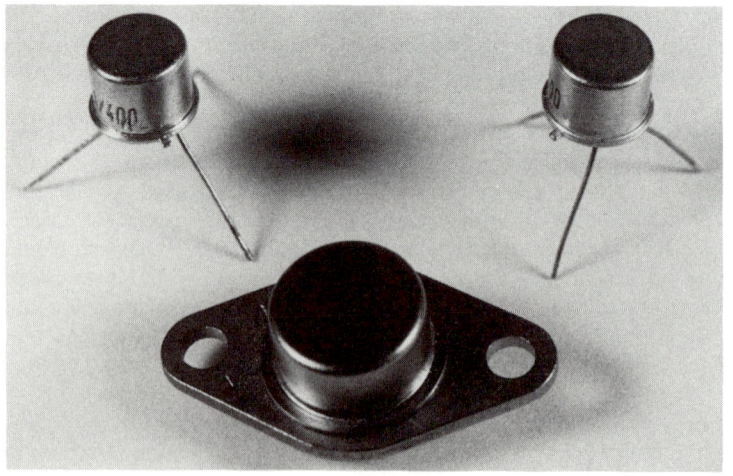

Figure 3.26 Two triacs and a thyristor; these are similar to the transistor in physical appearance

62

example is shown in Figure 3.26. Normally the device does not conduct between the anode and cathode terminals, but if a suitable positive bias current is fed through the gate—cathode terminals the device switches on and behaves in a similar manner to an ordinary silicon rectifier. Most types require a fairly high bias current of up to about 20 mA before they will trigger to the 'on' state, but quite high currents can be switched by these devices. Once triggered, a thyristor cannot be switched off again by removing the gate bias. It can only be switched off by reducing the anode-to-cathode current to a suitably low level (about 20 mA or less). Of course, if the device is used to handle only a small current, the 'hold-on' effect will not be produced, and on/off switching can be accomplished by means of the gate bias. Thyristors are sometimes sold under a particular type number, but often they are advertised by their two principal ratings: the maximum forward voltage that can be withstood, and the maximum current that can be controlled.

Light detector

The characteristics of a thyristor can be demonstrated using the light-detector circuit of Figure 3.27. This uses another component not described previously in this book: cadmium sulphide photoresistor

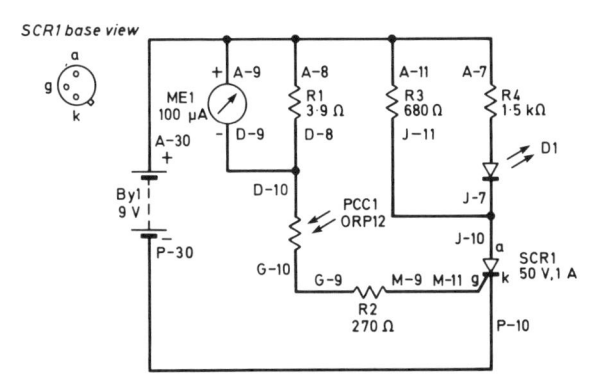

Figure 3.27 Light-detector circuit to demonstate the properties of a thyristor

PCC1. This form of resistor has a value that varies according to the light level received by its sensitive surface, and it has the modified circuit symbol shown in the diagram. The specified ORP12 device (included in Figure 3.28) has a value of more than 10 MΩ in total darkness, dropping to less than 100 ohms in very bright conditions. Its sensitive surface is the one opposite the leadout wires, incidentally.

If the photocell is in darkness when power is first applied to the circuit, little gate current will flow and SCR1 remains switched off. Subjecting PCC1 to a moderate or bright light provides SCR1 with a

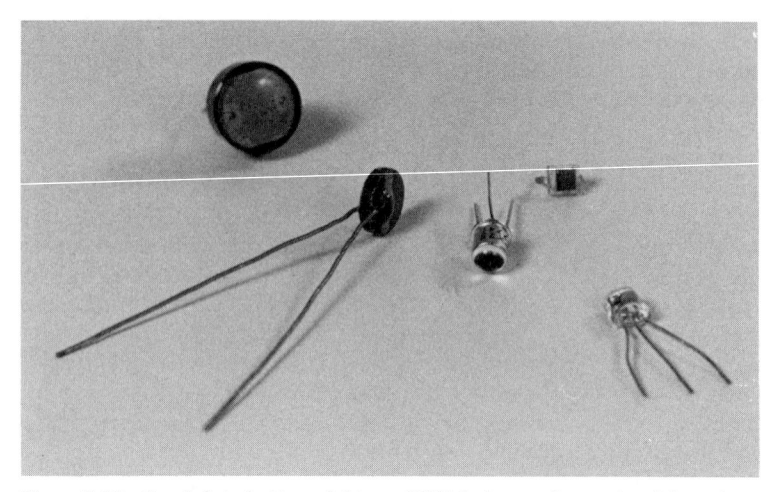

Figure 3.28 Top left to bottom right: an ORP12 photoresistor, a bead thermistor (see Chapter 4), a BPX25 phototransistor, a photodiode, and a 2N5777 Darlington-pair phototransistor

trigger current and it switches to the 'on' state. D1 lights up and ME1 indicates the increased gate current (the latter being limited to a safe level by R2). R3 ensures that the anode current of SCR1 is high enough to produce the hold-on effect, and so D1 remains on even if PCC1 is returned to darkness and the gate current falls back to its original level.

Triacs

A triac has the circuit symbol shown in Figure 3.25(b), and examples are included in Figure 3.26. It is very similar to a thyristor. The main difference is that when in the 'on' state it will conduct in either direction. Furthermore, it can be triggered by either a positive or a negative gate bias. It is used to control a.c. loads.

Photosensitive semiconductors

Semiconductor devices can be designed to act as photocells, and a number of different types are available. A *photodiode*, with the circuit symbol shown in Figure 3.25(c), has a reverse resistance that decreases with increasing light level. It can also be used as a voltaic cell (to generate a voltage from the received light). In either case photodiodes are relatively insensitive, but respond very rapidly to changes in light level.

A *phototransistor* has the circuit symbol of Figure 3.25(d). This type of component has a leakage current that increases with increasing light level. It is more sensitive than a photodiode, but not quite as fast in operation. Darlington-pair phototransistors are also available, and have a high level of sensitivity comparable to that of a cadmium sulphide cell, together with a reasonably high operating speed (cadmium sulphide cells are very slow in operation). Examples of these components are shown in Figure 3.28.

Field-effect transistors

There are actually several types of field-effect transistor (f.e.t.) available, the type in most common use probably being the junction-gate type (JUGFET). This has the circuit symbols of Figure 3.29(a) and (b). Its

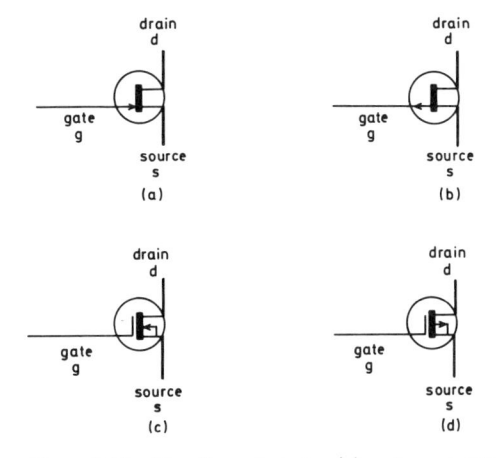

Figure 3.29 Circuit symbols for (a) n-channel JUGFET (f.e.t. equivalent of npn bipolar device); (b) p-channel JUGFET (f.e.t. equivalent of pnp bipolar device); (c) n-channel MOSFET; (d) p-channel MOSFET

three terminals are called the *drain, gate* and *source*, and are equivalent to the collector, base and emitter (respectively) of a bipolar transistor.

A JUGFET is very different to a bipolar device in that it has an input impedance of about 1000 MΩ, and it is the gate voltage rather than current that determines the level of drain—source conduction. Also, it is what is known as a *depletion-mode* device, which means that it is most conductive with a drain-to-source voltage of zero. It must be reverse biased for use in linear amplifiers. This is normally achieved using the arrangement shown in Figure 3.30. Here the gate is biased to the negative rail potential by R1. The drain-to-source current also flows through R3, producing a voltage across this resistor, and making the

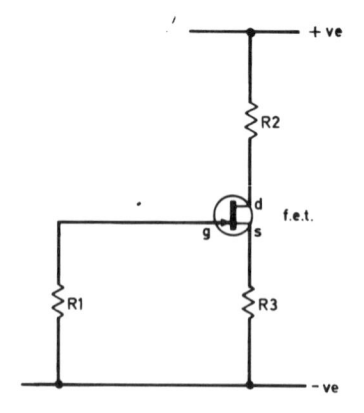

Figure 3.30 Normal method of generating reverse bias for a depletion-type f.e.t.

gate negative with respect to the source. This gives the required reverse bias.

Another type of f.e.t. is the MOSFET (metal oxide semiconductor f.e.t.), which has the circuit symbols of Figure 3.29(c) and (d). A depletion-mode MOSFET is biased in much the same way as a JUGFET, and has an even higher input impedance of about 1 000 000 MΩ. There are also *enhancement-mode* MOSFETs, which require a forward bias for linear amplifying applications, giving circuit configurations similar to those used for bipolar devices.

Additional components for Chapter 3

Resistors (all miniature ⅛, ¼ or ⅓ watt, 5% tolerance)

1 Ω	22 kΩ
3.9 Ω	39 kΩ
270 Ω	47 kΩ
360 Ω	1.8 MΩ
680 Ω	3.3 MΩ
1 kΩ	3.9 MΩ
3.9 kΩ	8.2 MΩ
4.7 kΩ (2 off)	10 MΩ
10 kΩ	ORP12 CDS photoresistor

Capacitors
4.7 μF 10 V electrolytic (2 off)
10 μF 10 V electrolytic (2 off)

Semiconductors
BC109 transistor (3 off)
BC169 transistor

BC179 transistor
(The above transistors sometimes have a suffix letter to indicate their gain ranking: 'A' types have the lowest gain, 'B' types have medium gain, and 'C' types the highest gain. All types are suitable for these circuits whether they have a suffix letter or not, and regardless of what suffix letter happens to be present.)
6.8 V zener diode (e.g. BZY88C6V8)
50 V 1 A thyristor (higher voltage types are suitable)

Miscellaneous
Crystal earphone (these are normally fitted with a 3.5 mm jack plug, and can be connected to the circuit demonstrator unit via a 3.5 mm jack socket to which a couple of single-strand leads about 75 mm long have been connected).

4

Operational amplifiers

Operational amplifiers are primarily designed to carry out mathematical operations in analogue computing, but they are actually used in a very wide range of applications. They have rather unusual characteristics which enable them to be used in d.c. amplifying applications, although they can be adapted to amplification of a.c. signals, and are frequently used in this way.

This chapter introduces three basic 'op amp' circuits, with experiments for you to demonstrate each of them. The additional components needed are listed on page 76.

It is possible to build an operational amplifier using discrete components (i.e. it can be assembled from individual transistors, resistors, etc.) but it is normal practice to employ an *integrated circuit* (i.c.) designed for this role. In somewhat oversimplified terms, an integrated circuit is a chip of silicon that has been processed to produce various components (transistors, diodes, resistors, and capacitors can all be formed), plus the interconnections necessary to combine these components into a circuit to perform the required function. This enables quite complex circuits to be packaged into a small encapsulation, and there is usually a considerable cost saving by using an i.c. instead of a discrete-component circuit of similar performance. Figure 4.1 shows a variety of i.c. encapsulations.

When dealing with i.c.s it is a perfectly viable approach to think of them as electronic building blocks having particular characteristics, and not to bother about the detailed workings of the internal circuitry. Indeed, there are so many i.c.s in common use that there is little alternative. The internal circuits of many devices are not published anyway.

Figure 4.1 Top row: an audio power i.c. with built-on heatsink, and a ZN414 i.c. Bottom row: a CD4001BE i.c. in conductive foam protective packaging, a 555 timer i.c. and a 741C op-amp i.c.

An operational amplifier has the circuit symbol of Figure 4.2. As can be seen from this, there are two inputs to the device: the non-inverting (+) input and the inverting (−) input. The output voltage of an operational amplifier is equal to the difference between the two input voltages multiplied by the voltage gain of the device. In theory an operational amplifier should have an infinite voltage gain; this is not achieved in practice of course, although practical circuits do have

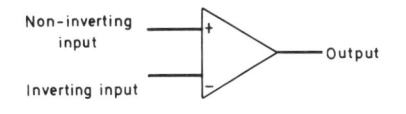

Figure 4.2 Circuit symbol for an operational amplifier

very high levels of gain. The d.c. voltage gain of the standard 741C device, for example, is typically some 200 000 times (20 000 mininum)! Thus only a minute voltage difference across the inputs will send the output fully positive or fully negative. It is important to remember that the gain of an operational amplifier is purposely progressively reduced at frequencies above a few hertz, in order to prevent instability. The gain of the 741C device at 10 kHz, for instance, is only about 100 times.

Comparator

The output of an operational amplifier goes positive if the non-inverting input is at a higher potential than the inverting input, and negative if the comparative input states are reversed. This can be demonstrated using the circuit of Figure 4.3. The inverting input is biased to about half the supply voltage by R1 and R2. With VR1 adjusted for the full supply potential at the non-inverting input, the latter will therefore be at the higher potential and the output of IC1 will be at virtually the full positive supply voltage. Adjusting VR1 for a progressively lower slider voltage causes the non-inverting input voltage to drop below the

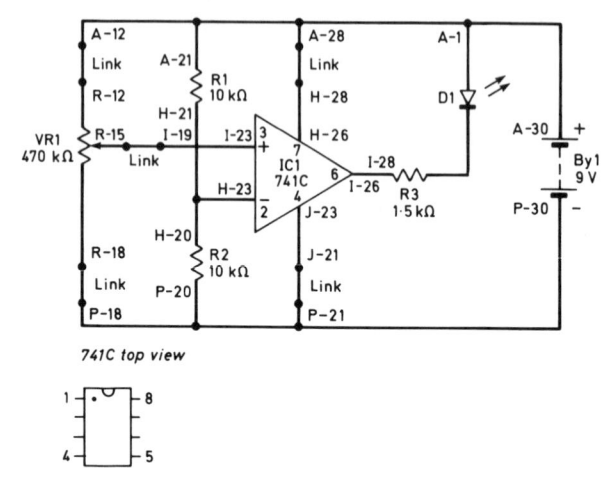

Figure 4.3 Circuit to demonstrate the op-amp as a comparator

inverting input potential when the control knob of VR1 is at about the centre of its adjustment range. This causes the output of IC1 to drop to within a volt or so of the negative supply rail, and this is indicated by D1 being switched on. By carefully adjusting VR1 to balance the two input voltages it may be possible to obtain an intermediate output voltage (indicated by D1 lighting up at less than full brightness), but there will only be a very narrow range of settings that produce such an output, if it is possible at all.

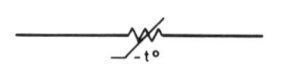

Figure 4.4 Circuit symbol for a thermistor (negative-temperature-coefficient type); this has a resistance that falls with increasing temperature, and rises with decreasing temperature

In this mode of operation the operational amplifier is working as a *voltage comparator* (often just called a comparator), and there are many applications for this type of circuit. For example, if R1 is replaced by a thermistor (type VA1066S is suitable), the circuit will operate as a temperature alarm. A thermistor has the circuit symbol shown in Figure 4.4, and one is included in Figure 3.28. A normal negative-temperature-coefficient type has a resistance value that *decreases* as the temperature of the component *rises*. If VR1 is adjusted for the lowest slider voltage that does not cause D1 to switch on, touching the thermistor causes it to be heated and its resistance to fall. This increases the voltage at the inverting input above the voltage at the non-inverting input, which results in D1 switching on to indicate the rise in temperature.

In practice VR1 would be adjusted to a setting that caused D1 to be switched on when the thermistor exceeded some particular temperature of importance. By using the output signal to control a heating element or refrigeration unit instead of an LED, this arrangement can be used as the basis of a high-accuracy thermostat.

Inverting amplifier

When used as an amplifier an operational amplifier can be used in either of two modes, and we shall consider the inverting mode first. A circuit to demonstrate the properties of the inverting-mode amplifier is given in Figure 4.5; the set-up on the demonstrator unit is shown in Figure

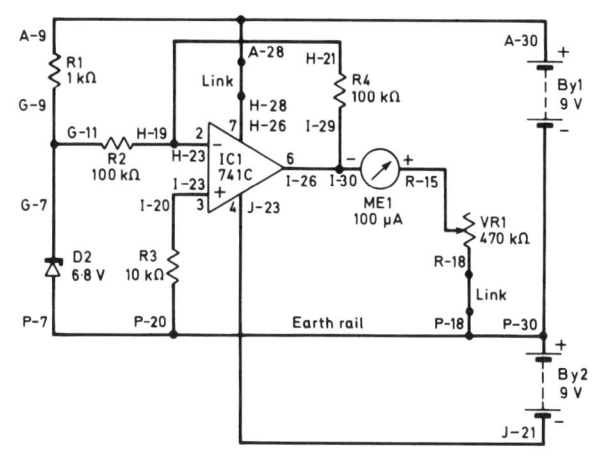

Figure 4.5 Circuit to demonstrate the action of an op-amp inverting amplifier; it also functions as a resistance meter

4.6. An unusual feature of this type of circuit is the use of dual balanced positive and negative supplies, By1 providing the former and By2 the latter. The output of the amplifier is able to swing positive or negative of the central earth rail. The non-inverting input of the operational amplifier is biased to this rail by R3.

A negative-feedback network consisting of two resistors is connected between the output and inverting input (R4), and the input signal and

Figure 4.6 Circuit shown in Figure 4.5 set up on the demonstrator unit

the inverting input (R2). If we ignore R1 and D2 for the moment, this biases the output to the earth-rail voltage. This occurs because a higher output voltage would, for instance, take the inverting input above the non-inverting input potential because of the coupling through R4. This would unbalance the input voltages and send the output negative to a point that balanced them. In theory the operational amplifier has an infinite input resistance (and in practice the input resistance is very high), so there is no voltage drop across R4. The output therefore assumes the earth-rail potential in order to balance the input voltages.

With R1 and D2 in circuit, an input voltage of 6.8 V is applied to the amplifier. This takes the inverting input positive, but the feedback action again tries to balance the input voltages, and this is achieved with an output potential of −6.8 V. The potential divider action across R4 and R2 then gives the required 0 V at the inverting input. This is indicated by a positive deflection on ME1.

If VR1 is adjusted for f.s.d. of the meter and R4 is then changed to a 10 kΩ component (disconnect both supplies before making this alteration), ME1 will read one-tenth f.s.d. when the supplies are restored. This is because the output now only needs to go 0.68 V negative in order to give 0 V at the inverting input, owing to the potential-divider action across R2 and R4. Thus the voltage gain of the circuit is controlled by the values of these two resistors. The voltage gain is actually equal to R4 divided by R2, and this is called the 'closed loop' gain. The innate voltage gain of the operational amplifier is called the 'open loop' gain.

The input impedance of the amplifier is equal to the value of R2 since the inverting input is maintained at earth potential by the feedback action, and what is termed a 'virtual earth' is formed at this input. Thus the input signal is effectively coupled across R2, and the input impedance is equal to its value.

Note that if R4 is replaced with various resistors having values of between a few kilohms and 100 kΩ, the meter reading in μA will always be equal to the value of the resistor in kilohms. In other words the circuit functions as a resistance meter!

Non-inverting amplifier

The other amplifying mode for an operational amplifier is the non-inverting mode. A circuit to demonstrate the action of this configuration is given in Figure 4.7. As before, the non-inverting input is biased to the earth-rail by a resistor (R2), but this time the input signal is coupled to

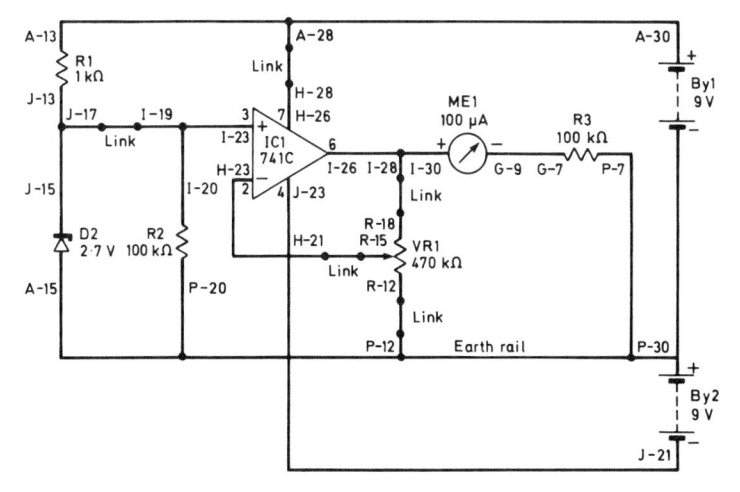

Figure 4.7 Circuit to demonstrate the action of an op-amp non-inverting amplifier

this input. A negative-feedback network is connected between the output and the inverting input, and one of the resistors is connected directly between these two points. In the circuit of Figure 4.7 this resistance is formed by the resistance between VR1's slider and upper track connection. The second resistor connects between the inverting input and the earth rail; in this circuit it is formed by the resistance between VR1's slider and lower track connection.

If we assume for the moment that the link wire between points J-17 and I-19 is absent, and there is no input signal for the amplifier, the negative-feedback action will bias the output to the earth-rail potential in much the same way as it did in the previous circuit. Any drift in the output voltage would unbalance the input voltage, causing the drift to be corrected.

If the slider of VR1 is taken to the top of its track, and the link wire is now connected into circuit, ME1 will indicate an output voltage of about +2.7 V (i.e. the same as the input voltage). This occurs because the negative-feedback action tries to balance the input voltages. With the non-inverting input at +2.7 V the output must go positive by the same amount in order to achieve this. If VR1's slider is taken (say) half-way down its track, then the output voltage will increase to +5.4 V, because double the output voltage is then needed in order to balance the input voltages owing to the potential-divider action of VR1. Thus, moving VR1's slider down its track increases the output voltage by increasing the voltage gain of the circuit.

The voltage gain is equal to $(R_1 + R_2)/R_2$ where R_1 is the resistance between the output and inverting input, and R_2 is the resistance between the inverting input and earth. As the input impedance of an operational amplifier is infinite (in theory anyway), the input impedance of the circuit is equal to the value of the non-inverting input's bias resistor.

Although positive input signals are used in the circuits of Figures 4.5 and 4.7, the inputs (and the outputs) can be of either polarity. With the inverting mode the output is always of the opposite polarity to the input signal, but with the non-inverting configuration they are always of the same polarity.

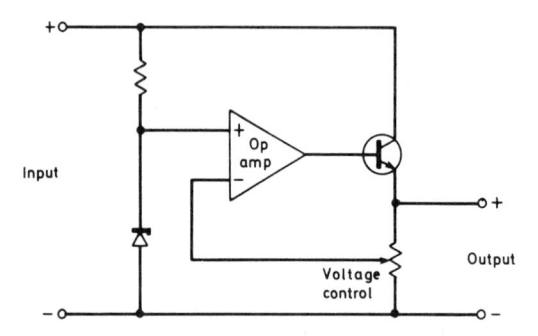

Figure 4.8 The Figure 4.5 configuration is often used as the basis of variable-voltage power supplies, with a discrete emitter-follower output stage being used in the manner shown here to give a high enough output current

A common application for a circuit of the type shown in Figure 4.7 is as a variable-voltage power supply, although it is usually necessary to add a discrete common-emitter output stage (as shown in Figure 4.8) in order to obtain a high enough output current, as most operational amplifier outputs can only supply a few milliamps. The zener diode provides a stable reference voltage, and the negative-feedback action ensures that the output voltage is accurately maintained at the factor of this potential, which is set using the output voltage control.

Offset null control

With practical amplifying circuits there is usually a small voltage across the inputs under quiescent conditions, although in theory they should be at precisely the same voltage. This is called the *input offset voltage*, and it can be a problem in some applications since the output will drift

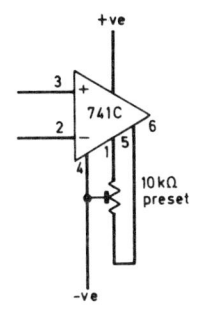

Figure 4.9 Method of adding an offset null control to a 741C device

away from the earth potential by an amount equal to the offset voltage multiplied by the closed-loop gain of the amplifier. Most operational amplifiers have provision for an 'offset null' control that can be adjusted for zero output voltage despite the input offset voltage. The 741C and many similar devices use the offset null arrangement of Figure 4.9.

Compensation capacitor

As mentioned earlier, the gain of an operational amplifier is reduced at high frequencies to prevent instability. This is usually achieved using a single capacitor, which is an internal component of many devices including the 741C. Some devices need an external 'compensation capacitor', as it is called; an example is the 748C i.c., which is virtually identical to the 741C in other respects. In unity-gain circuits there is

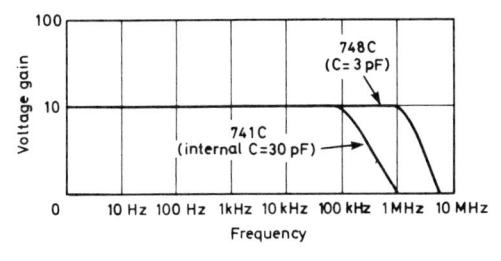

Figure 4.10 At closed-loop gains of more than unity it is possible to obtain a better high-frequency response with an externally compensated device than with a comparable internally compensated one

75

no advantage in using an externally compensated device, because the optimum compensation-capacitor value is then the same as that used in an internally compensated amplifier. At higher closed-loop gains it is possible to use a lower value than that used in internally compensated devices, and thus types requiring external compensation can achieve improved high-frequency response. This is shown in Figure 4.10, which gives the frequency response graph for a 741C used at a gain of ten times, and a 748C used at the same gain with the optimum compensation-capacitor value.

Of course, operational amplifiers are mainly used in d.c. and low-frequency circuits where obtaining the best possible high-frequency response is of little or no importance, and internally compensated devices are by far the more popular of the two types.

Additional components for Chapter 4

Resistor (miniature $^1/_8$, $^1/_4$ or $^1/_3$ watt, 5% tolerance)
10 kΩ

Semiconductors
VA1066S thermistor
741C integrated circuit, 8 pin DIL (dual-in-line) version
2.7 V zener diode (e.g. BZY88C2V7)

5
Oscillator and radio circuits

An oscillator is a circuit that automatically generates an a.c. or a varying d.c. signal. There are a great many types of oscillator, but this chapter describes just a few representative ones. Eight more experiments are included for you to try; the additional components needed are listed on pages 89—90.

Relaxation oscillator

There are several types of component that can be used as the active element of a relaxation oscillator, and probably the device most commonly used in this application is the *unijunction transistor*. A simple unijunction relaxation-oscillator circuit is shown in Figure 5.1. It must be stressed that a unijunction transistor has characteristics that are totally different from bipolar and f.e.t. transistors. It is not suitable for use in amplifying applications, and is normally only employed in relaxation-oscillator circuits.

The three terminals of a unijunction transistor (or u.j.t.) are called the *emitter, base 1,* and *base 2*. The device exhibits a resistance of a few kilohms between its base 1 and base 2 terminals, so in the circuit of Figure 5.1 there is a quiescent current of about 1—2 mA through R2, Tr1's B1—B2 resistance, and R3. There is a very high input impedance at the emitter (at least several megohms), and C1 is free to charge up via R1 and VR1. This it does until it achieves a charge potential of about two-thirds of the supply potential, at which point the u.j.t. 'fires'.

This is brought about by an internal regenerative action of the device, causing the input impedance at the emitter to drop to a low level, and the B1—B2 resistance to fall to only about half its normal figure. This

77

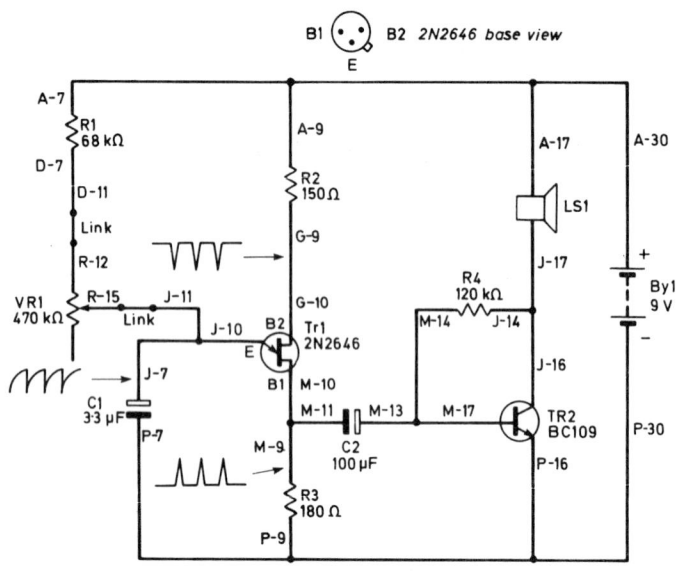

Figure 5.1 Relaxation oscillator using a unijunction transistor (Tr1); the circuit functions as an electronic metronome

causes the charge on C1 to be quickly leaked away, and the voltages on R2 and R3 to increase to about twice their previous amounts. When C1 has been largely discharged Tr1 reverts to its original state, enabling C1 to start charging up once again. When C1 has charged to a sufficient potential the u.j.t. 'fires' once again, and this whole process continually repeats itself. This gives a series of positive pulses across R3, negative pulses across R2, and a form of what is called a *sawtooth waveform* (for obvious reasons) across C1. In this circuit the positive pulses are coupled by C2 to a common-emitter amplifier using Tr2. Each time a pulse is received, Tr2 sends a burst of current through LS1, causing it to emit a clicking sound.

The frequency of oscillation depends upon the values of the timing resistance (R1 plus VR1) and the timing capacitance (C1), and is roughly equal to $1/CR$, with C in μF, R in megohms, and the answer in hertz. In theory this circuit has a frequency range that varies from about 4.5 Hz with VR1 at minimum resistance to approximately 0.56 Hz at maximum resistance, and the circuit will function as an electronic metronome giving results comparable to those of a conventional mechanical instrument. In practice this type of circuit is rather unpredictable with regard to frequency range, partly owing to the rather high tolerances of VR1 and C1 (typically ±20% and +50, −20% respectively), partly to the fact that the trigger voltage of Tr1 will vary considerably from one device to another. Thus it may be necessary to alter the value of C1 in order to obtain a suitable frequency range.

Transformer feedback

An amplifier can be made to oscillate by coupling the output to the input, but only if the following conditions are met:
- The feedback must be positive and not negative (i.e. if the output is positive-going, the feedback should provide an input signal that assists the output to go more positive, rather than trying to send it negative).
- There must be at least unity voltage gain through the amplifier and feedback components.
- The feedback provided should be a.c. and not d.c. (the latter would cause the output to latch in the fully positive or negative state).

It is not possible to obtain oscillation simply by coupling the output of a common-emitter stage back to its input, because the two are out of phase, and negative feedback would be provided. The same is true of an emitter-follower stage; although positive feedback would be obtained, the voltage gain of slightly less than unity would not support oscillation. A common-base stage provides voltage gain, and its input and output are in phase. However, a simple feedback arrangement would not support oscillation since the output would be loaded to such an extent by the input that the voltage gain would drop below unity.

In order to obtain oscillation from a single-transistor amplifier it is necessary to have a feedback circuit that will provide a suitable match between the output and input of the amplifier. The easiest way of achieving this match is to use a transformer, because it can provide a voltage or current step-up, as well as phase inversion if necessary. Figure 5.2 shows the circuit of a simple oscillator that uses a common-emitter amplifier with the feedback obtained via a transformer.

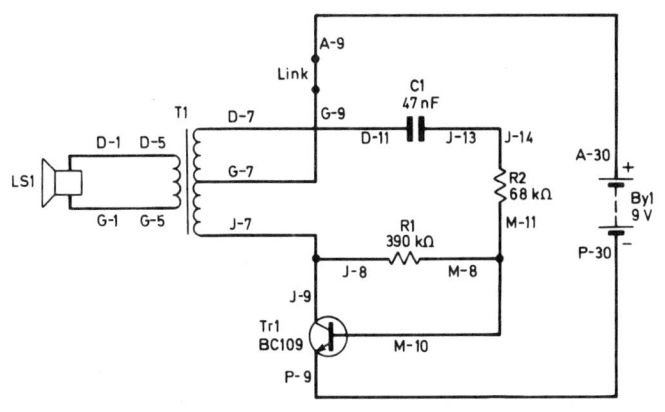

Figure 5.2 An oscillator circuit that uses a transformer to give in-phase feedback over a common-emitter amplifier

Tr1 is biased by R1, and the lower half of T1's primary forms its collector load. The upper half of the primary effectively forms a secondary winding connected out of phase with the lower section. When power is applied to the circuit the collector voltage of Tr1 begins to rise. This gives a negative-going signal from the uppermost connection of T1; this signal is coupled by C1 and R2 to Tr1 base, where it has the effect of holding Tr1 in the off state. The collector voltage therefore rises to the full supply potential. There is then no further coupling through C1 since the collector voltage becomes static, and R1 begins to bias Tr1 on, causing its collector voltage to drop. This gives a positive-going signal at the upper terminal of T1, and this signal is coupled to Tr1 base by C1 and R2. This causes Tr1 to conduct more heavily, resulting in a further fall in its collector voltage, and a continuation of the positive-going signal applied to its base. Eventually Tr1 becomes saturated (conducting as heavily as it can), with Tr1 collector at little more than zero volts. The coupling through C1 then ceases as the collector voltage becomes static. R1 is unable to supply the base current needed to keep Tr1 in saturation, and Tr1 therefore begins to switch off. Its collector voltage rises, giving a negative-going signal from the upper connection of T1, which eventually results in Tr1 being switched off. This brings the circuit back to its original state. The process then starts again from the beginning, giving continuous oscillation.

The frequency of oscillation is determined by a number of factors, including the values of C1 and R2, the input impedance of Tr1, and the characteristics of T1. Using the specified components the circuit should oscillate at a frequency of about 400 Hz, producing an audio tone from the loudspeaker which is fed from the true secondary winding of T1.

There are a number of other oscillator configurations that use transformers, but although this type of oscillator was once popular for simple applications such as Morse-practice oscillators and continuity testers, they are only infrequently used nowadays.

Phase-shift oscillator

In the circuit of Figure 5.3 (shown set up in Figure 5.4), a common-emitter amplifier is made to oscillate by applying the feedback through a three-section phase-shift network. These networks consist of C2–R1, C1–R2, and C3–Tr1 input impedance. Each phase-shift network delays the signal by less than 90 degrees, and at one frequency they will provide a delay of 60 degrees (this assumes that the phase-shift networks have identical component values, which is standard practice in circuits of this type). There is then a total phase shift of 180 degrees, so that the signal is inverted through the networks and positive feedback is provided. In

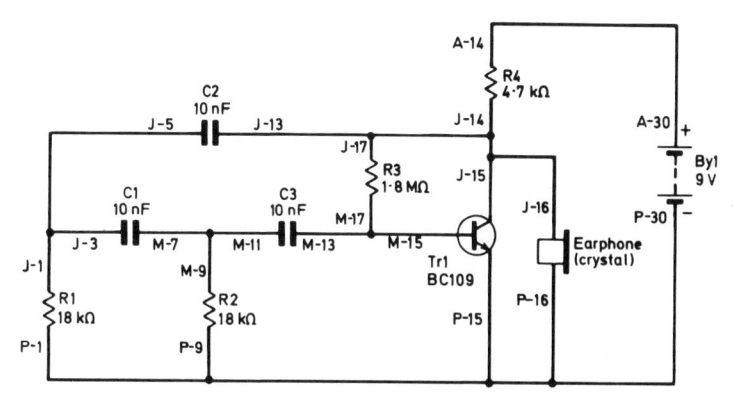

Figure 5.3 Circuit to demonstrate a phase-shift oscillator

Figure 5.4 Circuit shown in Figure 5.3 set up on the demonstrator unit

practice the phase-shift networks will not have precisely the same component values as each other, but this merely results in some providing less than 60 degrees phase shift, and others providing more. There will be 180 degrees of shift through the entire network at a certain frequency, and the circuit oscillates at this frequency. In theory the common-emitter stage needs a gain of 29 times to compensate for the losses through the phase-shift networks, although it is slightly more than this in practice, owing to the minor variations in the component values of each network resulting in higher losses than the theoretical value. The gain of Tr1 should be more than adequate, though, giving strong and reliable oscillation.

Sine wave

The output from a phase-shift oscillator is a reasonably pure sine wave — the waveshape of Figure 3.16(a). A sine wave contains just one frequency component, whereas any other waveshape contains a basic or *fundamental* frequency plus some combination of other frequencies, usually *harmonics* (multiples of the fundamental frequency). By listening to the output of the phase-shift oscillator using the earphone it will probably be possible to hear the strong fundamental signal plus weaker higher-frequency harmonics. The circuit does not give a completely pure sine wave, because it would be necessary for the gain of Tr1 to be reduced to a level just adequate to sustain oscillation in order to achieve this. In practice there would still be small impurities on the output, owing to the variations in the gain of Tr1 as its collector current varied.

A phase-shift oscillator (also known as a 'dippy' oscillator) operates at a theoretical frequency of

$$\frac{1}{2\pi\sqrt{(6CR)}}$$

or a frequency of about 400 Hz with the component values shown in Figure 5.3.

Astable multivibrator

This is a popular form of oscillator, and uses the configuration shown in the circuit diagram of Figure 5.5. It really just consists of a common-emitter amplifier based on Tr1, capacitively coupled to a second

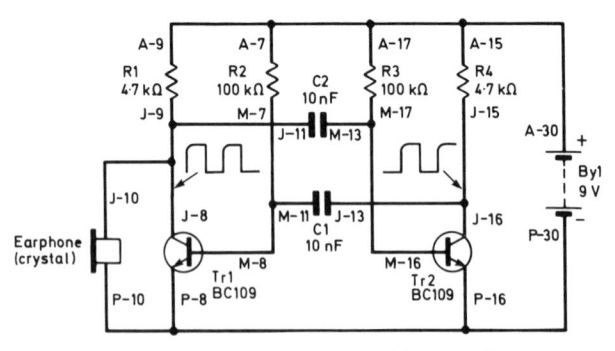

Figure 5.5 Circuit to demonstrate an astable multivibrator

common-emitter stage using Tr2, with positive feedback provided through C1. The feedback is positive because both stages invert the signal, giving an in-phase relationship to the input and output of the amplifier as a whole. The high gain of the circuit gives an output waveshape that is roughly square, and is rich in harmonics. The signals at Tr1 and Tr2 collectors are 180 degrees out of phase. The frequency of oscillation is approximately equal to $1/1.4CR$, where C is the value of C1 (C2 should have the same value) and R is the value of R2 (R3 should have the same value). The circuit of Figure 5.5 oscillates at about 714 Hz.

Integrated circuits are often used in oscillator circuits, and Figure 5.6 shows an astable circuit based on a type 555 i.c. Although this is usually referred to as an astable, it is actually a form of relaxation oscillator. C2 charges via R1 and R2 until it achieves a potential equal to two-thirds of the supply voltage. The i.c. is then triggered and discharges C2 via R2 into pin 7 of the device. This continues until the charge on C2 falls to one-third of the supply voltage, and then the i.c. reverts to its original state, enabling C2 to start charging up once again. The circuit continually oscillates in this manner, producing a roughly triangular waveform across C2.

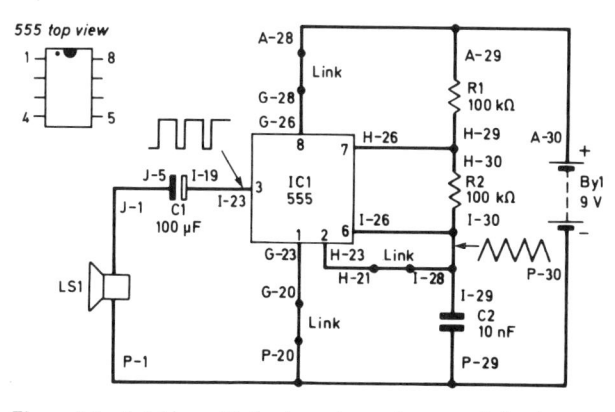

Figure 5.6 Astable multivibrator using an integrated circuit

The main output of the device is at pin 3. It assumes a high voltage when C2 is charging, but only a fraction of a volt when it is being discharged. This gives a rectangular waveform of low impedance (the 555 has a complementary class B output stage). Quite a loud audio tone will therefore be emitted from the loudspeaker, so this type of circuit is ideal for use in projects where an audio alarm signal is required. The frequency of oscillation is approximately equal to

$$\frac{1.44}{(R_1 + 2R_2)C}$$

or about 480 Hz with the values shown in Figure 5.6.

83

Tuned circuits

A tuned circuit usually consists simply of a capacitor and an inductor connected in parallel, as shown in Figure 5.7(a). We have seen earlier how a capacitor can store an electric charge; in a sense, an inductor is

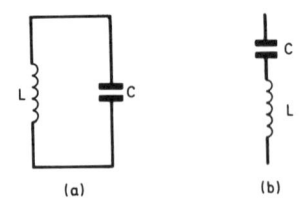

Figure 5.7 Tuned circuit: (a) parallel, (b) series

also capable of holding a charge. In fact it gives up the charge as soon as power is removed from it, but thereby causes the magnetic field around the inductor to collapse, inducing a voltage in the component. If power is fed into a tuned circuit, therefore, when the power source is disconnected a voltage develops across the inductor and charges the capacitor. The capacitor then discharges through the inductor, producing a collapsing field and a voltage which charges the capacitor once more. The voltage produced across the inductor is opposite in polarity to the collapsing input voltage provided by the capacitor, and this gives an alternating current in the tuned circuit. In theory the oscillations in the tuned circuit continue for ever, but in practice leakage currents flow through the capacitor, and (of greater significance) there are losses due to the resistance in the wiring and inductor. This causes the oscillations to die away quickly.

The frequency of the oscillations depends on the values of the inductor and capacitor. The larger the capacitor, the longer it takes to discharge and the lower the frequency generated. Increasing the value of the inductor also slows the discharge of the capacitor and the build-up of the reverse charge, giving lower-frequency oscillations. The frequency of the oscillations is called the 'resonant frequency', and is equal to

$$\frac{1}{2\pi\sqrt{(LC)}}$$

Figure 5.7(b) shows an alternative form of tuned circuit, the series type, but this is rarely used in practice.

Tuned-circuit oscillator

Tuned circuits are often used in oscillator circuits, mainly in high-frequency circuits where they tend to give better frequency stability than C–R types. These circuits are basically the same as ordinary transformer feedback types, but the primary and/or secondary windings of the transformer have a parallel-connected capacitor to produce a tuned circuit. The circuit of a simple oscillator using a tuned circuit is shown in Figure 5.8.

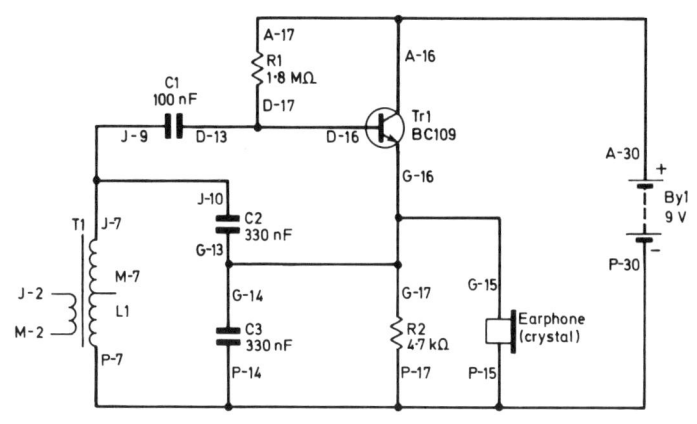

Figure 5.8 An oscillator circuit using a tuned circuit (L1 is the large winding of LT700 transformer)

Tr1 is used in the emitter-follower mode, and the tuned circuit consists of the large winding of T1 plus the series capacitance of C2 and C3. The tuned circuit is coupled to the input of Tr1 by C1. The output from Tr1 emitter is coupled to the junction of C2 and C3, which forms what is called a 'capacitive tapping'. This in effect forms a centre tap on the tuned circuit, and a 2:1 voltage step-up is obtained through the tuned circuit, giving sufficient feedback to produce oscillation. (A step-up can be obtained in the same way by applying an input signal to a tapping on a transformer; this arrangement is termed a 'single-wound' transformer. An ordinary transformer having two windings is called a 'double-wound' type.)

The frequency of oscillation is equal to the resonant frequency of the tuned circuit, since it is only close to this frequency that the circuit provides an efficient coupling of the feedback. This can be demonstrated by increasing C2 and C3 to 1 μF, and then reducing them to 100 nF. This will cause the output frequency first to decrease owing to the lower resonant frequency, and then to increase owing to the higher resonant frequency.

Crystal set

High-frequency signals of between a few tens of kilohertz and thousands of megahertz can be coupled to an aerial and radiated in the form of a magnetic and electrical field, or 'radio waves'. An aerial (which can consist of just a length of wire) can be used to pick up the electrical field and produce a current. In effect, as the waves travel past the aerial they produce a varying potential across the wire, with a consequent alternating current flow.

The radio signal can be modulated in some way so that it carries a useful signal (speech, music, etc.) from the transmitter to the receiver. The form of modulation used by medium- and long-wave broadcast stations is *amplitude modulation* (a.m.). This superimposes the audio signal on the radio frequency (r.f.) 'carrier' signal in the manner shown in Figure 5.9. The receiver must select the desired transmission's

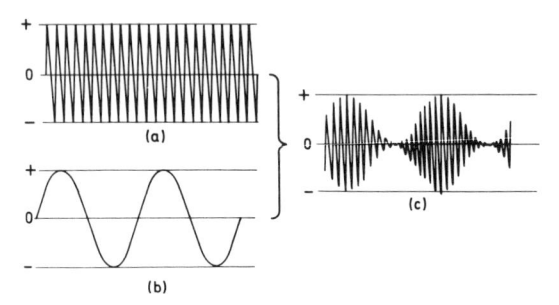

Figure 5.9 An r.f. signal, (a), can be amplitude modulated by an audio signal, (b), to produce the waveform (c)

carrier frequency, reject carriers on other frequencies, and convert the received r.f. signal into the required modulating audio signal. The simplest type of receiver that can do this is the crystal set, and Figure 5.10 gives the circuit of a simple receiver of this type.

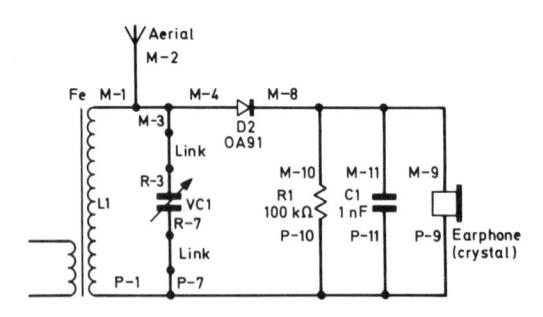

Figure 5.10 Circuit diagram of a simple crystal radio set

If the set is tried out in practice, the aerial can consist of about 10 to 30 metres of p.v.c.-insulated connecting wire, which should be positioned as high as possible. L1 and VC1 form a tuned circuit, which can be adjusted to resonate at any frequency on the medium waveband using VC1. L1 is wound on a ferrite rod, and it may be necessary to experiment with the position of the coil on the rod in order to obtain the correct inductance value and frequency coverage. (Note that there is a smaller winding over one end of L1; this is simply ignored as it is not required in this application.) The purpose of the tuned circuit is to filter out the required transmission from the multitude of other stations. At its resonant frequency there is a high-impedance path through the tuned circuit. In theory it has an infinite impedance at resonance, since the signal fed into it sets up continuous oscillation. In practice there are losses in the circuit, and some of the input signal at the resonant frequency is absorbed. Away from the resonant frequency there is a comparatively low-impedance path to earth through VC1 and L1. Thus, except for signals close to the resonant frequency of the tuned circuit, input signals from the aerial are shunted to earth, and so the desired station can be selected by adjusting VC1 for the appropriate resonant frequency.

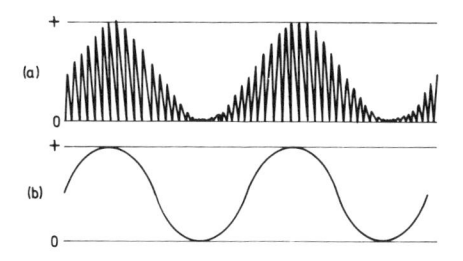

Figure 5.11 (a) The detector diode rectifies the r.f. input signal to give an output of this type; (b) after the r.f. signal has been smoothed, the original modulating a.f. waveform is left

D2 simply rectifies the carrier to give an output of the type shown in Figure 5.11(a). The time constant of R1 and C1 is made long enough to smooth the r.f. half-cycles from D2, but not so long as to smooth out the variations in average amplitude that are produced by the modulation. This gives the waveform of Figure 5.11(b) across the earphone, and this is of course the original modulating audio signal. D2, R1, and C1 form a single a.m. 'detector' or 'demodulator'.

One problem with a crystal set is its lack of 'selectivity'. In other words it is not very good at picking out just the required station, and where there are several strong stations close together they may well all be received at once! This is partly because L1 does not have a high 'Q' value. The Q of a coil is a measure of how close or otherwise it comes to theoretical perfection. The higher the Q the more efficient the inductor and the better the selectivity obtained in this application. The situation is made worse by the loading of the detector, which considerably reduces the effective Q of L1. Taking less signal from the tuned circuit would give better selectivity, but there would probably then be inadequate volume.

Radio set using an integrated circuit

The selectivity problem can be overcome by coupling the output of the circuit to an amplifier so that a lower level of loading will give adequate sensitivity. A circuit of this type is given in Figure 5.12, based on the

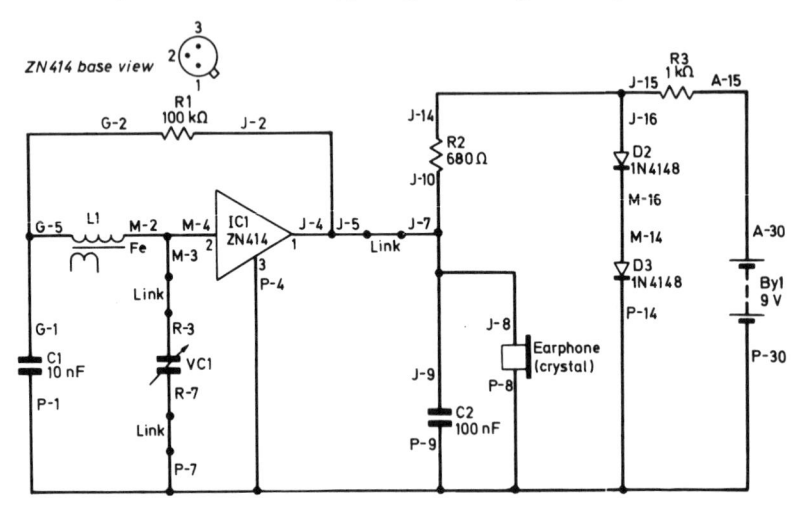

Figure 5.12 Improved radio circuit using an i.c. to give gain control and reduced aerial loading

ZN414 radio i.c. This device has a high input impedance that does not significantly load the tuned circuit. The ZN414 is biased by discrete resistor R1, and this provides the bias current through L1 so that it does not shunt the input impedance of the circuit. C1 provides an earth-return path for the earthy end of L1. IC1 has an internal detector circuit that only requires discrete load resistor R2 and r.f. filter capacitor C2 for correct operation. Note that the positive supply for IC1 is obtained only through R2, and there is no direct positive-rail connection to the device. A supply potential of only about 1.3 V is needed; this is derived from the 9 V supply using a sort of zener stabiliser arrangement using R3 and forward-biased silicon diodes D1 and D2.

Results using this circuit should be far superior to those obtained with the crystal set with regard to both selectivity and sensitivity. An external aerial is not necessary because the magnetic field of the radio signal induces a sufficient voltage in L1 to give good results. The circuit is designed so that when strong signals are received the d.c. voltage at the output of IC1 reduces, and gives a reduction in the gain of the circuit. This effect can be clearly shown by connecting the 100 μA meter between the output of IC1 and the negative supply rail via a 39 kΩ resistor. There will be a small but noticeable dip in the meter reading when a strong station is tuned in. This gives a simple form of automatic gain control (a.g.c.), which prevents strong signals from overloading the receiver. It also helps to give a fairly constant volume level from the received stations even though they give signals of greatly differing strengths. Another beneficial effect is that it helps to combat fading on stations that are prone to this problem.

Receivers (such as the two described above) that detect the r.f. signal at its transmitted frequency are called 'tuned radio frequency' (t.r.f.) sets. Most radios actually use a more complicated arrangement called the superhet circuit. This converts the received signal to an 'intermediate frequency' (i.f.), where it is considerably amplified and processed by several tuned circuits prior to detection. Excellent sensitivity and selectivity can be obtained in this way, as can a highly effective a.g.c. circuit as well. There are other forms of modulation apart from the standard a.m. type, such as frequency modulation (f.m.) which is commonly used on the 'very high frequency' (v.h.f.) bands. However, it is not possible to consider these in more detail here.

Additional components for Chapter 5

Resistors (all miniature $^1/_8$, $^1/_4$ or $^1/_3$ watt, 5% tolerance)
180 Ω
18 kΩ (2 off)
68 kΩ
120 kΩ

Capacitors
1 nF plastic foil or ceramic
10 nF plastic foil (3 off)
47 nF plastic foil
100 nF plastic foil
330 nF plastic foil (2 off)
3.3 μF 10 V electrolytic

Semiconductors
555 integrated circuit
2N2646 unijunction transistor

OA91 germanium diode
1N4148 silicon diode (2 off)
ZN414 integrated circuit

Miscellaneous
Denco MW5FR ferrite aerial (or any other m.w. ferrite aerial)
Wire for aerial

6

Pulse and logic circuits

Electronic circuits can be designed not only for *linear* applications, where the input and output signals are continuously variable between certain limits; many circuits are designed to generate and process *pulse* signals. Circuits of the latter type have only two stable states in which the inputs and outputs can exist: 'high' and 'low'. The former means at or close to the positive supply rail voltage, and is also known as 'logic 1'. The low state means at or close to the negative supply rail voltage, and is also called 'logic 0'. It has been assumed here that a normal negative-earth system is used. For positive-earth circuits high (logic 1) is equal to virtually the negative-rail potential and low (logic 0) is virtually equal to the positive-rail voltage.

This final chapter includes five circuit-demonstration experiments; the additional components needed are listed on page 102.

Bistable multivibrator

This is one of the simplest switching circuits and has the basic configuration of Figure 6.1. When power is initially applied to the circuit both transistors will begin to switch on, as there is a path from each base to the positive supply via two different sets of two resistors. Owing to the component tolerances and stray circuit capacitances, one transistor will turn on faster than the other, and for the sake of this example we will assume that Tr1 is the faster of the two. This keeps Tr1's collector voltage low, starving Tr2 of base current through R2 and sending its collector voltage high. This increases the base current to Tr1, helping to keep its collector voltage low and cut off the bias to Tr2. This regenerative action ceases with Tr1 saturated and Tr2 switched off.

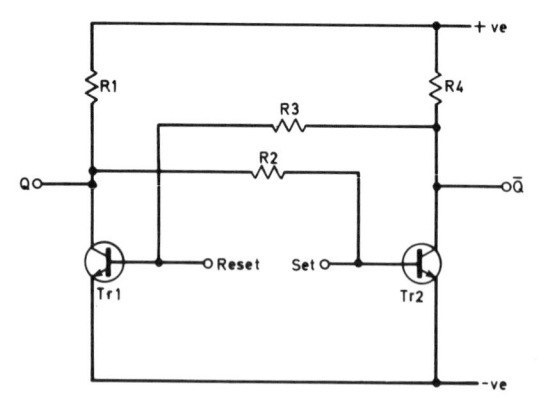

Figure 6.1 Basic bistable multivibrator configuration

The circuit can be set to the opposite state, with Tr2 saturated and Tr1 switched off, by applying a positive signal to Tr2's base. This switches on Tr2, causing its collector voltage to fall and remove the bias formerly fed to Tr1 by way of R3. Tr1 collector then rises to the high state, holding Tr2 in the on state owing to the bias current via R2 when the input signal is removed. (The same effect could be obtained by taking Tr1 base negative.)

The circuit can be reset to its original state by briefly applying a positive signal to Tr1 base, so that it switches back on and Tr2 is once again cut off. Thus it is possible to continually trigger the circuit from one state to the other; in effect it provides a sort of simple memory action, retaining the last state it was forced into. Outputs are available

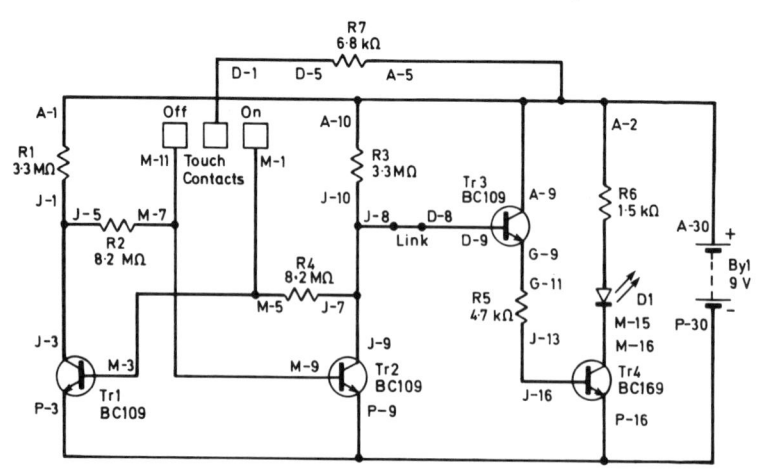

Figure 6.2 Bistable multivibrator (formed by Tr1, Tr2, R1, R2, R3, R4) as part of a touch-switch circuit

at the collectors of both transistors, and they always assume opposite states. Complementary outputs of this type are normally called Q and Q̄ (the latter meaning 'not Q').

Although possible applications for pulse circuits are perhaps less obvious than for linear circuits, their range of uses is at least as wide. A bistable multivibrator can, for example, be used as the basis of a simple touch-switch circuit such as that shown in Figure 6.2. If Tr1 collector is in the high state, emitter-follower stage Tr3 will supply a heavy base current to common-emitter transistor Tr4, biasing it hard on and switching on the load. In this case the load is an LED indicator, but it could be any small 9 V d.c. device such as a transistor radio. Touching the 'off' touch contacts causes a current to flow through R7 and the operator's finger into Tr2 base, switching on Tr2 and altering the output states of the bistable. Tr2 collector therefore falls to a low level, switching off Tr3, Tr4 and the load. Operating the 'on' contacts causes Tr1 to be switched on, altering the output states of the bistable and switching the load back on again.

If you construct the touch switch on the circuit demonstrator unit, the leadout wires to the touch contacts can be wired to one side of a three-way terminal block. Short 6BA screws can be fitted into the other side of the block and will form suitable touch contacts to test the circuit.

Monostable multivibrator

As the name suggests, a monostable has just one stable state; if it is triggered into its secondary state it will only remain there for some predetermined time before automatically reverting to its stable state. In other words a monostable produces an output pulse of fixed duration from a brief input pulse, regardless of the duration of the input pulse.

Monostables can be produced from discrete components, but these days it is far more common to employ i.c.s intended for this application. Figure 6.3 gives the circuit diagram of a simple electronic timer unit using a 555 i.c. in the monostable mode. (See also Figure 6.4.) The device is triggered by applying a negative pulse to pin 2; this is provided by C1 and R2 when power is first applied to the circuit, as C1 will initially be uncharged and take pin 2 to the negative supply potential. When the device is triggered the output at pin 3 goes to the high state, and therefore LED indicator D1 does not switch on. Normally C2 is held in a discharged state by an internal transistor of the i.c., but this is removed when the device is triggered, enabling C2 to commence charging. The 555 has a voltage-detector circuit that senses when the charge on C2 reaches two-thirds of the supply voltage, and resets the circuit to its normal state at this point. This sends the output low and

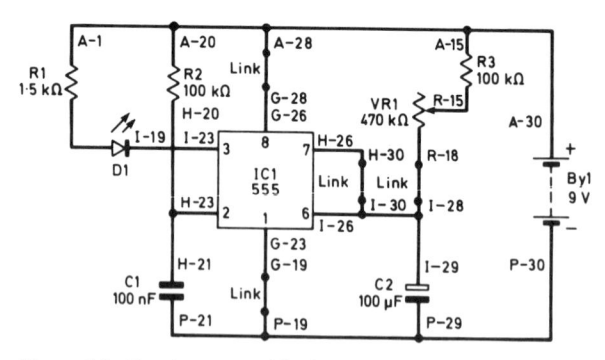

Figure 6.3 Simple monostable timer circuit based on a 555 i.c.

Figure 6.4 Circuit shown in Figure 6.3 set up on the demonstrator unit

switches on D1. The circuit can be started on another timing run either by disconnecting and reconnecting the supply, or by momentarily short-circuiting C1. The output pulse length of this circuit is approximately equal to 1.1CR, and is variable over a nominal range of 11 seconds to just over one minute using the specified values.

Note that if the trigger input is taken low, and kept low beyond the end of the timing period, the output will not return to the low state until the input pulse ceases. Thus the circuit only functions properly if the input pulse is shorter than the required output pulse length. This type of circuit is sometimes called a 'pulse stretcher', or a 'retriggerable monostable'. There is a non-retriggerable type where the output pulse length is totally independent of the input pulse length. Also, some monostables require a positive trigger pulse and/or produce a negative output pulse. Many monostables actually have complementary Q and Q̄ outputs, and some monostable i.c.s can be used in any monostable mode.

Schmitt trigger

A Schmitt trigger is used to produce an output that is either high or low from an input signal that does not switch cleanly and rapidly between these two states. Normally the output is low, but if the input is taken above a certain threshold voltage the output almost instantly switches to the high state. If the input is then taken below this threshold level the output remains high. It is necessary to further reduce the input potential, below a second threshold level, before the output will revert to the low state. This effect is known as 'hysteresis', and removes the possibility of the output falsely switching between states producing numerous output pulses, which could easily occur if there was a single threshold level and the input potential hovered close to it. The spurious operation could result either from instability in the Schmitt trigger itself or from noise on the input signal. In either case the hysteresis eliminates the problem.

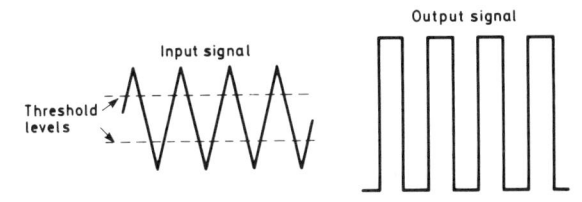

Figure 6.5 A Schmitt trigger can be used to produce a rectangular output waveform from non-rectangular input signals

Probably the main use of Schmitt trigger circuits is to process a sine wave, triangular wave, or some similar waveform, to produce a rectangular output signal that can be used to operate pulse and logic circuitry reliably. Figure 6.5 shows how this transformation is achieved.

The circuit of Figure 6.6 shows a common Schmitt trigger arrangement using two transistors, although i.c.s are increasingly used in this application. With a very low voltage at VR1's slider, Tr1 will be cut off and Tr2 will be biased hard on by the base current it receives through R1 and R2. An LED indicator (D1) has been included in the collector circuit of Tr2, and this comes on of course. If the voltage at VR1 slider is gradually increased, eventually there will be a sufficient bias potential to begin biasing Tr1 into the 'on' state. This results in a fall of Tr1's

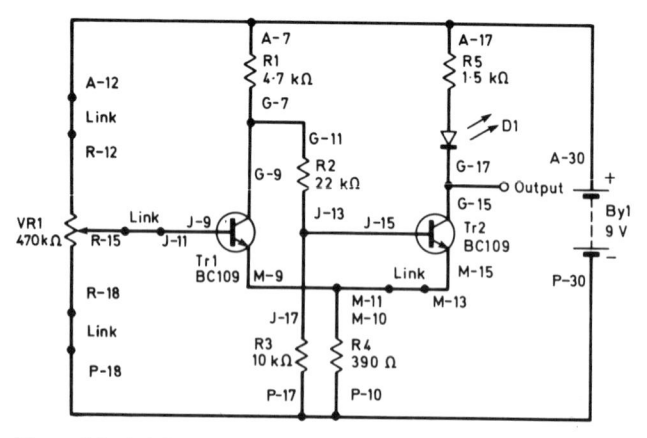

Figure 6.6 A Schmitt trigger circuit

collector potential, thus reducing the bias fed to Tr2. This in turn results in a fall in the emitter and collector currents of Tr2, with the voltage developed across R4 falling in consequence. Tr1 is therefore biased harder into conduction since the fall in its emitter voltage increases the base—emitter bias voltage. This gives a regenerative action, which causes Tr1 to switch on quickly and Tr2 to switch off just as rapidly. This causes the output at Tr2 collector to change almost instantly from the low to the high state, which will be shown by D1 switching off.

With the reduced voltage across R4, the voltage at VR1's slider must now be substantially reduced before Tr1 begins to switch off, and a reverse regenerative action causes the output to trigger from the high state to the low state. This gives the required hysteresis.

Gates

It is not possible to consider many of the various types of logic circuit here, but as an introduction to logic techniques the simplest of logic circuit elements, the *gates*, will be covered. The simplest type of gate is the 'NOT' type, or 'inverter' as it is more commonly known. An inverter can be produced using one transistor and two resistors in the arrangement of Figure 6.7. If the input is taken to logic 0, Tr1 is switched off and the output assumes the high state (as will be registered by the voltmeter). Taking the input to logic 1 biases Tr1 into saturation, and the output goes to logic 0. In other words the output always assumes the opposite logic state to the input. A table known as a 'truth table' is often used to show the function performed by a logic device; the truth table for a NOT gate is given in Figure 6.8. This particular table is of

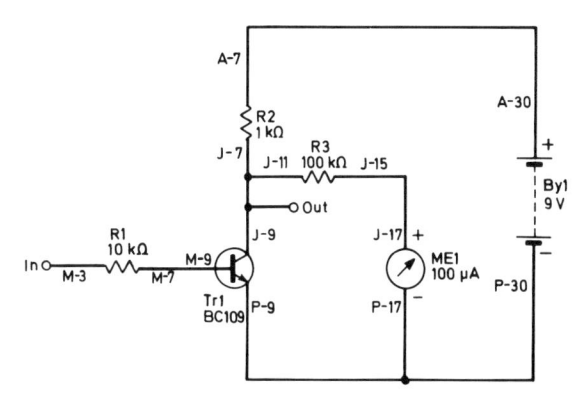

Figure 6.7 A single-transistor NOT gate circuit

Input	Output
1	0
0	1

Figure 6.8 Truth table for a NOT gate

little practical use since the action of an inverter is quite apparent without the aid of the table. However, with more complex devices a truth table can be of considerable help.

Most gates have two or more inputs and one output, the logic state at the output depending upon what combination of logic states is present at the inputs. The four common types of gate are the AND, OR, NAND and NOR. These have two or more inputs, there being no upper limit on the number of inputs.

The output of a two-input AND gate is high if inputs 1 *and* 2 are high, but low for any other set of input states. For a three-input AND gate the output is high if inputs 1 *and* 2 *and* 3 are high, but is low for any other combination. In fact the output of an AND gate is high if all the inputs are high, and is low with any other set of input states, regardless of the actual number of inputs.

A two-input OR gate has a high output state if input 1 *or* input 2 *or* both are high, and a low output state if they are not. A three-input OR gate has a high output condition if input 1 *or* input 2 *or* input 3 is high, *or* if more than one input is high, but a low output if the inputs are all low. The output of an OR gate is always high if one or more of the inputs is high, and low if all the inputs are low, regardless of the actual number of inputs.

A NAND gate can be regarded as an AND gate with a NOT gate added at the output. Similarly, a NOR gate functions like an OR gate that has a NOT gate added at the output.

97

The truth tables for two-input AND, OR, NAND and NOR gates are shown in Figure 6.9. The five types of gate considered here have the circuit symbols shown in Figure 6.10, although unfortunately there are several alternatives to these in common use. Integrated-circuit gates are available at very low prices, and it is normal to use these rather than discrete components wired to perform the required logic function.

AND				OR		
Input 1	*Input 2*	*Output*	*Input 1*	*Input 2*	*Output*
0	0	0	0	0	0
0	1	0	0	1	1
1	0	0	1	0	1
1	1	1	1	1	1

NAND				NOR		
Input 1	*Input 2*	*Output*	*Input 1*	*Input 2*	*Output*
0	0	1	0	0	1
0	1	1	0	1	0
1	0	1	1	0	0
1	1	0	1	1	0

Figure 6.9 Truth tables for two-input AND, OR, NAND and NOR gates

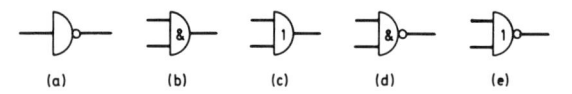

Figure 6.10 Circuit symbols for: (a) NOT gate (b) AND gate, (c) OR gate, (d) NAND gate, (e) NOR gate

Those who would like to prove the action of one or two gates in practice can use the circuit of Figure 6.11. This can be used to check the performance of one of the four gates in the CD4001 quad two-input NOR device, or the CD4011 quad two-input NAND device. The LED indicator comes on if the output is high, and is unlit if the output is low. The inputs are coupled to the appropriate supply lines via link wires to provide the required input logic states. The two specified i.c.s are in the CMOS (complementary metal oxide semiconductor) range of logic devices, and it should perhaps be pointed out that owing to the extremely high input impedances of these devices (about 1.5 million megohms) they can be damaged by high static voltages. However, they incorporate protection circuitry, and although it is possible that they

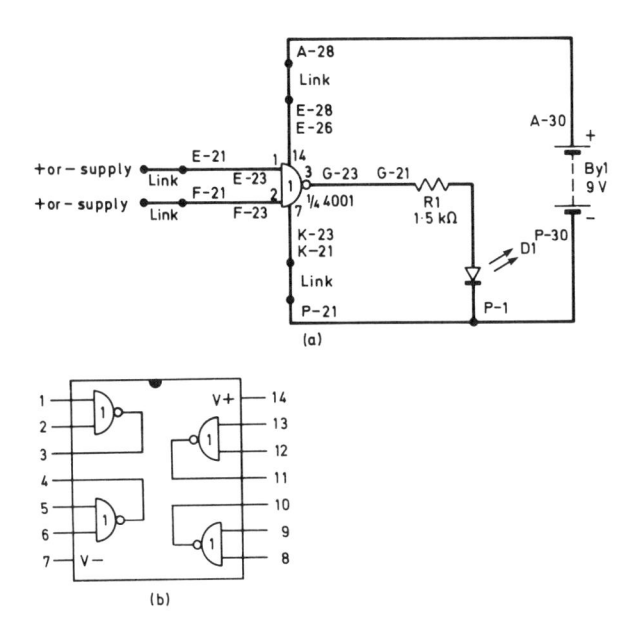

Figure 6.11 (a) Circuit to demonstrate the action of a NOR (or NAND) gate; (b) pin arrangement of the CD4001 i.c. (the 4011 uses the same general arrangement)

could be damaged by static charges while carrying out the experiments, it is highly unlikely that this would occur. In normal use these devices would be left in their protective packaging until it was time for them to be plugged into circuit. They would be the last devices to be connected into circuit and would be handled as little as possible.

Note that NOR and NAND gates can be made to operate as inverters simply by wiring the inputs together, and this technique is often used in circuits where spare gates are available and some inverters are required.

There is obviously a limit to the number of inputs that can be fed from the output of a logic circuit, and in the case of CMOS devices at least 50 inputs can be driven from each output. This figure is called the 'fanout'.

Gates are sometimes used as the only active components in applications such as simple electronic games and alarm systems, but they often play a relatively unimportant role in complex digital equipment such as timers, frequency counters, etc.

A typical use for a logic gate is shown in Figure 6.12, which is the circuit diagram of a simple burglar alarm system. The normally open (n.o.) switches could be the well known trigger mats that produce a short circuit between two contacts when someone steps on one of them. The normally closed (n.c.) switches might be door and window types that open if one of the doors and windows to which they are fitted is

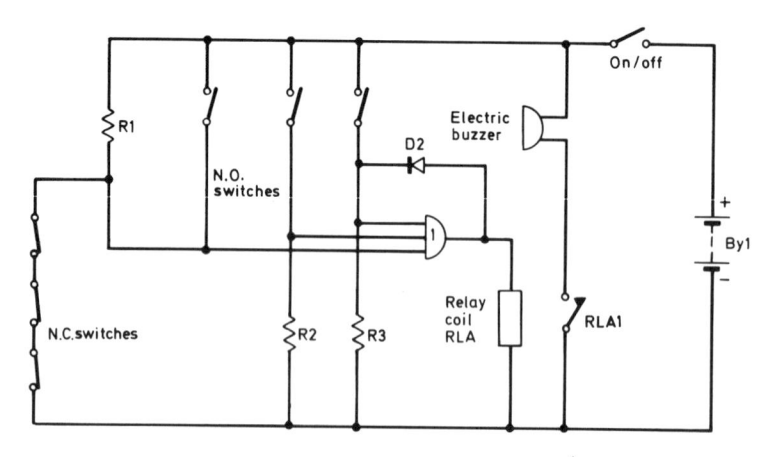

Figure 6.12 Simple burglar alarm circuit incorporating a three-input OR logic gate

opened. As the circuit stands, input 3 is held in the low state by the normally closed switches, while inputs 1 and 2 are taken low by R3 and R2 respectively. The gate is an OR type, and as none of its inputs are high the output assumes the low state. No power is therefore supplied to the load connected at the output, which is a relay. A relay is simply a switch that is activated by an electromagnet, and in this case the relay contacts are a set of the normally open type. These close when the relay coil is energised, providing power to the electric buzzer. The circuit symbols used for relays are explained in Figure 6.13.

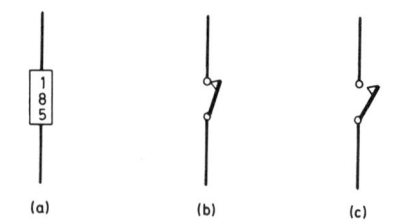

Figure 6.13 (a) Circuit symbol for a relay coil; these are annoted RLA, RLB etc. (b) Circuit symbol for a 'normally closed' relay contact. Contacts for RLA are labelled RLA1, RLA2 etc; those for RLB are labelled RLB1, RLB2 etc.

If an intruder should trigger one of the switch mats, one of the normally open switches will close and take one of the gate's inputs high. This causes the output to go high, switching on the relay and causing the alarm buzzer to be activated. The same thing would happen if the intruder were to cause one of the normally closed switches to open, as input 3 of the gate would then be taken high by R1. Once the output has gone high it latches in that state because of the coupling between the output and input 1 through D2. The output holds input 1 in the high state, and with an input at this state the output must remain high. Thus it is only necessary to trigger the circuit momentarily in order to cause a continuous alarm signal to be generated.

Binary

Single logic outputs are fine for applications such as controlling equipment where an on/off switching action is all that is required, as in our burglar alarm example. However, much digital electronics is concerned with processing complex data, and this requires a number of logic lines to be used together. The patterns of 1s and 0s on a set of digital outputs can be used to represent numbers, letters of the alphabet, or just about anything you like. Obviously some method of coding is needed though.

Numbers are represented by digital outputs in binary form. With the binary numbering system only 1s and 0s are used, but in many other ways it is much the same as the ordinary decimal numbering system. With a decimal number like 123, the 3 is the number of units, the 2 is the number of tens, and the 1 is the number of hundreds. In other words, the columns represent units, tens (10 to the power of 1), hundreds (10 to the power of 2 or ten squared), thousands (ten to the power of three), and so on. With binary numbers the columns represent units, twos (2 to the power of 1), fours (2 squared), eights (2 cubed), sixteens (2 to the power of 4), etc. Like the decimal numbering system, any integer can be represented by a unique combination of figures. The drawback of the binary system is that it takes a lot of digits to represent even quite small values. Most digital systems operate using groups of eight bits (called a 'byte'), or multiples of eight bits. As 8 bits gives a range of just 0 to 255, any task that involves even moderately large numbers will require at least two bytes to be used together to enable these numbers to be accommodated. A lot of computers now work using basic 16 and 32 bit chunks of data (often termed 'words' and 'long-words' respectively). On the other hand, much digital electronics only requires eight bytes of data.

This method of handling data may seem to be rather a crude one, and I suppose in some ways it is, but it is the method almost universally adopted in electronic systems because it can be easily handled by electronic circuits. Although it is not very convenient for users of digital equipment, the equipment can be (and usually is) designed to provide some form of 'user-interface' that leaves anyone operating the equipment unaware of the fact that the equipment is using binary arithmetic. An ordinary pocket calculator is a good example of this. You press the buttons to enter data in what appears to you to be decimal form, and the answers are displayed in what also seems to be ordinary decimal form. In truth, the push buttons are providing

101

0 and 1 logic levels, and the segments of the display are on or off. Electronically the input and output signals are in standard digital logic 0 and logic 1 form. An encoder circuit is used to convert key presses into data in a form that the main processing circuit can understand and manage properly. Decoder circuits are then used to convert the processed data into a form that can derive some form of display that will give an ordinary decimal output.

Converting binary numbers to their equivalent decimal values is not a difficult mathematical task for humans. From right to left, the columns in an eight bit binary number represent 1s, 2s, 4s, 8s, 16s, 32s, 64s and 128s. Where there is a 1 in a column you simply add the appropriate figure to the total, or if there is a 0 that column contributes nothing to the total. As an example, the binary number 10001110 is equivalent to 142 in decimal, as demonstrated below:

```
128s    64s    32s    16s    8s    4s    2s    1s
1       0      0      0      1     1     1     0
128  +  0  +   0  +   0  +   8  +  4  +  2  +  0  =  142
```

BCD

The conversion from ordinary binary to decimal can also be achieved by electronic circuits, and computer systems often use this form of pure binary. Non-computer digital systems often utilise a variation on the binary system called BCD (binary coded decimal). This differs from standard binary in that it uses groups of four bits of data to represent decimal digits. Therefore, the binary number 10000001 (129 decimal) is 81 in BCD. The eight bits of the number really have to be considered as two separate groups of 4 bits (or 'nibbles' as they are sometimes termed). The two nibbles are 1000 and 0001, which are 8 and 1 respectively when they are converted into decimal.

Two points about BCD should be apparent. The range of values available using eight bits is just 0 to 99, and not the 0 to 255 of ordinary binary. It stems from this that some 8 bit binary numbers are not legal BCD values. Each nibble must never be higher than 9 (1001 in binary). Incrementing a nibble from 1001 should take it back to zero, and increment the next most significant digit by one. Circuits must be designed to work in standard binary or in BCD as the two systems are largely incompatible. Some digital integrated circuits have a mode control input that enables either ordinary binary or BCD operation to be selected, as required.

Hex

There is another method of numbering which is often encountered in electronics, and this is the hexadecimal system, or just 'hex' as it is more commonly known. Binary is convenient in that it gives a clear view of the actual logic levels present at the outputs of a circuit. It is inconvenient in that it tends to involve large numbers. Decimal is easier in that it gives more manageable numbers in a more familiar form, but it is much harder and slower when it is necessary to translate numbers into the actual logic levels present in the circuit. Hexadecimal attempts to give the best of both worlds by giving two digit representations of eight bit binary numbers, and making it easy to translate these numbers into binary codes if required.

I suppose that hexadecimal is really just an extension of the BCD system, and in our BCD example above, 10000001 in binary is 81 in BCD and is also 81 in hexadecimal. The difference between the two systems is that whereas each nibble in BCD can not be more than 9 (1001), in hexadecimal all sixteen binary codes are valid. This brings a problem in that there are only ten single digit numbers available (0 to 9), but for hexadecimal we need sixteen. The problem is solved by using the first six letters of the alphabet (A to F) to provide the six additional numbers. Thus, when a hexadecimal counter is incremented from 9 (1001) it goes to A (1010). The list provided below should clarify the way in which hexadecimal operates.

8 bit binary	Hexadecimal
00000000	00
00000001	01
00000010	02
00000011	03
00000100	04
00000101	05
00000110	06
00000111	07
00001000	08
00001001	09
00001010	0A
00001011	0B
00001100	0C
00001101	0D
00001110	0E
00001111	0F

```
00010000        10
00010001        11
00010010        12

11111101        FD
11111110        FE
11111111        FF
```

Counters

A typical application for simple binary or hex circuits would be a batch counter. This would most probably take the form shown in Figure 6.14, where some form of opto-electronic sensor detects objects passing on a conveyor-belt, and produces one pulse per object. The pulses are fed into a series of BCD counters. The first of these starts at 0, and increments to 9, after which it resets to zero on the next input pulse.

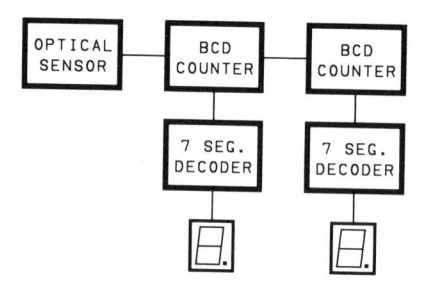

Figure 6.14 A batch counter using BCD counters

However, BCD counters almost invariably have a 'carry' output, and this provides a signal that is one tenth of the input frequency. In other words, the output stays low for five input pulses, then goes high for five input pulses, then goes low for a further five input pulses, and so on. This signal is coupled through to the next counter where the high to low transition after ten input pulses increments the second digit from 0 to 1. Thus the circuit provides a count from 0 to 99. By adding more counters in series it is possible to obtain a count as high as you like.

Numeric displays are virtually all of the seven segment variety these days. These have the familiar figure of eight pattern (Figure 6.15), with the segments identified by letters from A to G. I suppose that the name 'seven' segment display is not entirely accurate, as there are actually eight segments including the decimal point (DP) segment which is normally at the bottom right-hand corner.

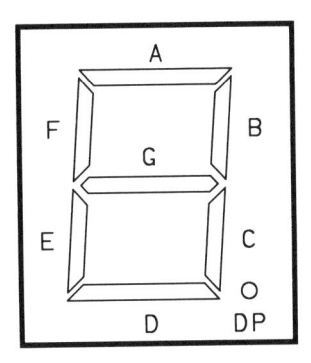

Figure 6.15 Seven segment display identification letters

The four outputs of a BCD counter can not correctly drive a seven segment display without the aid of a BCD to seven segment decoder circuit. These simply process the BCD input codes to provide seven outputs, one per display segment, and some BCD counters have built-in decoders. These are normally provided instead of the standard BCD outputs rather than as additional outputs.

A batch counter is just one example of numerous uses for this type of counting circuit. They are to be found in numerous pieces of digital equipment, including such things as frequency meters, timers, digital multimeters and digital thermometers. Unfortunately, BCD counter circuits are somewhat beyond the capabilities of the circuit demonstrator unit, but you might like to experiment with the simple binary counter circuit of Figure 6.16.

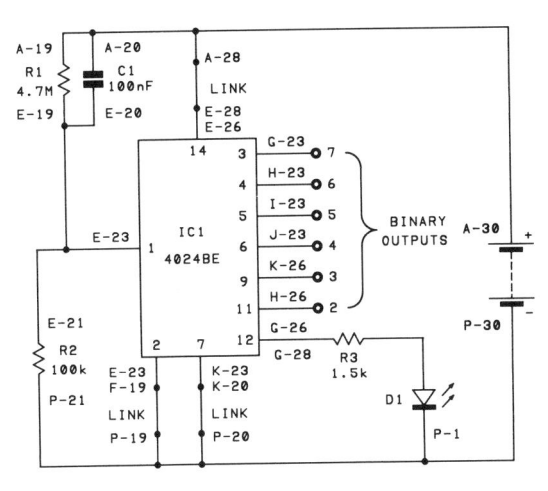

Figure 6.16 The basic binary counter/divider circuit

105

This uses a CMOS 4024BE counter, which is a seven stage binary type, or a 'ripple' counter as these are often called. As shown in the circuit diagram, LED indicator D1 is driven from output 1 of the counter. Start with one lead of R2 left unconnected. You should find that D1 can be switched on by connecting and disconnecting one lead of R2 to produce an input pulse. Producing another input pulse should switch it off, another one should switch it on again, and so on. If you try the same thing using output '2' the result should be much the same, but with two pulses needed per change of output state. Moving on to output '3' you should find that four pulses per change in output level are required, while output '4' will require eight pulses per change.

As you may have deduced from this, a binary counter is just a series of divide by two circuits. In fact devices such as the 4024BE are often used as dividers and not binary counters. For instance, where a highly accurate low frequency 'clock' signal is needed, it is standard practice to use a high frequency (quartz crystal controlled) oscillator plus a divider chain to reduce the final output frequency to the desired figure.

You will notice that two resistors and a capacitor are used at the input of the counter to generate the clock pulses, where a single resistor and a link-wire would seem to be sufficient. The additional components are needed to provide 'debouncing', which means removing the extra pulses that tend to be generated by mechanical switching. These additional pulses are not only generated by a very crude form of switching like the one used in this case, and even most high quality mechanical switches seem to suffer quite severely from this problem. This debouncing circuit operates by having C1 charge rapidly when R2 is connected into circuit, but it discharges only slowly through R1 when R2 is disconnected again. This tends to filter out any noise spikes, with the input staying low due to the charge on C2 during any spurious breaks in the switching action that should occur. When R2 is disconnected, the charge on C2 gradually subsides, and again removes any switching glitches. Some logic circuits are intolerant of slowly changing input levels, but this simple system of debouncing does not seem to give any problems with the 4024BE. Incidentally, the slow time constant of the circuit means that the maximum pulse frequency that can be generated is only about two per second.

There are binary dividers which have more than seven stages, but larger dividers can be produced simply by coupling the final output of one device to the clock input of the next, and adding as many devices in series as necessary. A problem with long divider chains in true counter (rather than divider) applications is that there is a small delay between each section receiving a pulse and responding to it. This type of counter is an 'asynchronous' type, and in high speed applications this delay, although it is usually less than a ten millionth of a second, can be significant. A 'synchronous' counter is one where the delays have been equalised so that all stages operate in unison, or appear to do so anyway.

As a final experiment, try connecting pin 2 to the positive supply rail and then reconnect it to the 0 volt supply rail again. Pin 2 is the 'reset' input, and this should have the effect of setting all the outputs to logic 0. Counters always have a reset input, and in normal use this is supplied with a pulse at switch-on so that the counters start at zero. If this is not done they will commence at a random number. This is not usually of importance in simple divider applications, but is unlikely to be acceptable in true counting applications. With circuits that use several counter devices, the reset terminals are usually connected together. They are then all reset to zero at switch-on, and are reset in unison by any reset pulses generated in the normal operation of the circuit.

Supply decoupling

Supply decoupling is necessary in many circuits, including both linear and logic types. As mentioned earlier, a battery has an internal resistance that causes its output voltage to fall to some degree as the current drain is increased. Mains power supplies also have a source impedance and suffer from the same problem, although to a much lesser extent with sophisticated stabilised types. This can result in changes in the current consumption of one stage causing variations in the supply voltage and consequent changes in the voltages present in other stages. This unwanted coupling between stages can cause malfunctions in logic circuits, with spurious input and output signals being generated. In linear circuits this stray coupling causes feedback from the output stage to an earlier stage or stages. If the feedback is of the negative variety it results in a loss of gain. If the feedback is positive it may well cause the

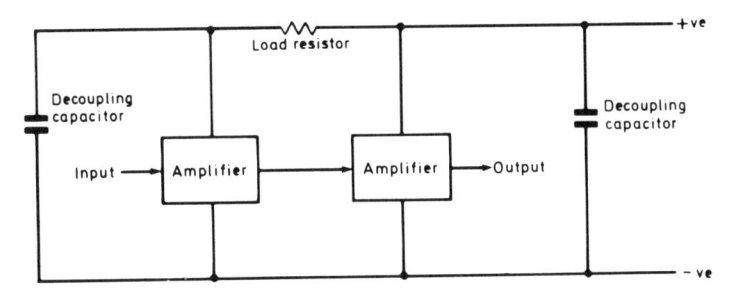

Figure 6.17 A method of supply decoupling often used in audio designs

107

circuit to break into oscillation, usually at a low frequency and termed 'motor boating'.

Supply decoupling often just consists of a capacitor connected across the supply lines to smooth out the variations in supply voltage and thus decouple the unwanted feedback. In high-frequency circuits a capacitor of between about 1 and 100 nF is normally adequate for this purpose, and a value of about 100 nF is typical in pulse circuits. In audio circuits the variations in supply voltages are comparatively slow and quite high currents may be involved, necessitating the use of fairly large decoupling-capacitor values of typically 100 μF or more. Sometimes, in order to give a large amount of decoupling with reasonably low decoupling-capacitor values, the arrangement of Figure 6.17 is used.

Additional components for Chapter 6

Resistors (all miniature 1/8, 1/4, or 1/3 watt, 5% tolerance)
390Ω
6.8kΩ
3.3MΩ
4.7MΩ
8.2MΩ

Semiconductors
CD4001BE or equivalent (i.e. any 4001 CMOS device)
CD4011BE or equivalent (i.e. any 4011 CMOS device)
CD4024BE or equivalent (i.e. any 4024 CMOS device)

Index

Aerial, 85–86
Alternating current (a.c.), 30–32
Ampere (amp), 12, 14, 15
Amplifiers,
   class B, 59–61
   common-base, 56–57
   common-collector (emitter
      follower), 55, 58
   common-emitter, 53, 54–55
   inverting, 71–73
   microphone, 57–58
   non-inverting, 73–74
Amplitude modulation (a.m.), 86
Anode, 38
Astable multivibrator, 82–83
Automatic gain control (a.g.c.), 89
'Avalanche' effect, 43, 44

Base (of transistor), 46, 77
Battery, 12, 30
BCD, 102
Biasing, 49–53, 59
Binary, 101
Bistable multivibrator, 91–93
Bootstrapping, 55–56
Bridge, 30
Bridge rectifier, 42–43
Burglar alarm, 99–101
Bypass capacitor, 56
Byte, 101

Capacitors, 25–33
Carrier, 86
Cathode, 38
Charge, 11
Charge time, 27–28
CMOS, 98, 99
Clipping, 50–51
Coder, 102
Coil (inductor), 33
Collector (of transistor), 46
Colour coding, resistor, 17–19
Comparator, 69–71
Compensation capacitor, 75–76
Conductors, 12–13
Coulomb, 12, 28
Counter, 104
Coupling, 31–32
Current flow, 11–17
Current meter, 15

Darlington pair, 59–60, 62, 65
Debouncing, 106
Decoder, 102
Decoupling, 101
Demodulator (detector), 87
Dielectric, 27
Diodes, 38–46
Direct coupling, 58
Direct current (d.c.), 30
Distortion, 61–62

109

Divider, 106
Drain (of transistor), 65

Electrolytic capacitor, 25—26
Electrons, 11—14
Emitter (of transistor), 46, 77

Farad, 28—29
Feedback,
 negative, 51, 52—53, 62, 71, 73, 79, 101
 positive, 79, 80, 82, 101
Ferrite aerial, 87
Field-effect transistor (f.e.t.), 65—66
Free electrons, 11—14
Frequency, 31
Full scale deflection (f.s.d.), 15
Fullwave rectification, 42—43
Fundamental frequency, 82

Gain, 47, 48, 55, 58, 69, 72, 74
Gate (of transistor), 63, 65
Gates, logic, 96—101
Germanium, 40, 47

Halfwave rectification, 41
Harmonics, 82
Henry, 33
Hertz, 31
Hexadecimal, 103
Hold-on current, 63
Hysteresis, 95

Impedance, 34, 53—54
Inductors, 33—37
Insulators, 12
Integrated circuit (i.c.), 68, 83, 88, 93, 98
Intermediate frequency (i.f.), 89
Inverter, 96
Inverting amplifier, 71—73

JUGFET, 65

Kirchhoff's law, 22—23

Leakage, 27, 47
Light detector, 63—64
Light-emitting diode (LED), 7, 9, 45—46
Linear potentiometer, 58—59

Logarithmic potentiometer, 58—59
Logic circuits, 96—101

Metronome, 78
Modulation, 86, 89
MOSFET, 66
Multivibrators,
 astable, 82—83
 bistable, 91—93
 monostable, 93—95

Non-inverting amplifier, 73—74

Offset null control, 75
Ohm, 13, 16
Ohm's law, 14—17
Operational amplifiers, 68—76
Oscillators, 77—85

Peak inverse voltage, 40
Peak-to-peak output voltage, 50
Phase, 42
Phase-shift oscillator, 80—81, 82
Photocells, 63, 64—65
Potential divider, 23, 52
Potentiometer, 24, 58—59
Power, 24—25
Primary winding, 34, 35
Pulse circuits, 91—96
Push-pull rectifier, 42

Relaxation oscillator, 77—78
Radio circuits, 85—89
Radio frequency (r.f.), 86
Reactance, 32—33, 34
Rectifiers, 38, 40—43
Relays, 100
Reset, 107
Resistance meter, 73
Resistors, 13—25
Resonant frequency, 84, 85, 87
Ripple, 41, 42
Ripple counter, 106

Sawtooth waveform, 78
Schmitt trigger, 95—96
Secondary winding, 34, 35
Seven segment display, 104
Silicon, 40, 47
Sine wave, 82

Smoothing, 41, 42
Source (of transistor), 65
Soldering, 5—7

Temperature alarm, 70—71
Thermal runaway, 61
Thermistor, 70—71
Thyristor (silicon controlled rectifier), 62—64
Time constants, 29
Touch switch, 93
Transformers, 34—37, 79
Transistors, 46—62, 65—66, 77
Triacs, 64
Truth tables, 96—98

Tuned circuits, 84—89

Unijunction transistor (u.j.t.), 77

Volt, 13
Voltage, 13—17, 21—23
Voltage comparator, 70
Voltage divider, 23, 35, 52
Voltage stabiliser, 43
Voltmeter, 21—22, 26—27, 28

Watt, 25
Words, 104

Zener diodes, 43—45